McDougal Littell Middle School

COURSE 2

Math

Larson Boswell Kanold Stiff

CHAPTER 1

Resource Book

The Resource Book contains a wide variety of blackline masters available for Chapter 1. The blacklines are organized by lesson. Included are support materials for the teacher as well as practice, activities, applications, and project resources.

McDougal Littell
A HOUGHTON MIFFLIN COMPANY
Evanston, Illinois • Boston • Dallas

Contributing Authors

The authors wish to thank the following individuals for their contributions to the Chapter 1 Resource Book.

Donna Foley
Rebecca Salmon Glus
Julie Groth
Mark Johnson
Jessica Pflueger
Joanne Ricci
Corinne M. Schaeffer

ISBN: 0-618-26864-2

3456789–MDO–07 06 05 04

Contents

CHAPTER 1 Number Sense, Patterns, and Algebraic Thinking

Contents

Contents

Descriptions of Resources

This Chapter Resource Book is organized by lessons within the chapter in order to make your planning easier. The following materials are provided:

Tips for New Teachers These teaching notes provide both new and experienced teachers with useful teaching tips for each lesson, including tips about common errors and inclusion.

Parents as Partners This guide helps parents contribute to student success by providing an overview of the chapter along with questions and activities for parents and students to work on together.

Lesson Plans and Lesson Plans for Block Scheduling This planning template helps teachers select the materials they will use to teach each lesson from among the variety of materials available for the lesson. The block-scheduling version provides additional information about pacing.

Activity Masters These blackline masters make it easier for students to record their work on selected activities in the Student Edition, or they provide alternative activities for selected lessons.

Technology Activities with Keystrokes Keystrokes for various models of calculators are provided for each Technology Activity in the Student Edition where appropriate, along with alternative Technology Activities for selected lessons.

Practice A, B, and C These exercises offer additional practice for the material in each lesson, including application problems. There are three levels of practice for each lesson: A (basic), B (average), and C (advanced).

Study Guide These two pages provide additional instruction, worked-out examples, and practice exercises covering the key concepts and vocabulary in each lesson.

Real-World Problem Solving These exercises offer mathematical problem-solving activities for the material in selected lessons in a real-world context.

Challenge Practice These exercises offer challenging practice on the mathematics of each lesson for advanced students.

Chapter Review Games and Activities This worksheet offers fun practice at the end of the chapter and provides an alternative way to review the chapter content in preparation for the Chapter Test.

Projects with Rubric These projects allow students to delve more deeply into a problem that applies the mathematics of the chapter. Teacher's notes and a 4-point rubric are included. The projects include a real-life project, a cooperative project, and an independent extra credit project.

Cumulative Practice These practice pages help students maintain skills from the current chapter and preceding chapters.

CHAPTER 1

Tips for New Teachers

For use with Chapter 1

Lesson 1.1

TEACHING TIP As students work with easier patterns such as those on page 5, encourage them to write the operation required to move from one number in a pattern to the next, such as +7 in Example 1. As students work with more complex patterns, it will be essential to approach numerical patterns this way.

COMMON ERROR Many students believe that a division pattern such as,

64, 32, 16, 8, . . .

will eventually yield 0 or a negative number. Show them that this is not the case as the next numbers in this pattern are

$4, 2, 1, \frac{1}{2}, \frac{1}{4}, \ldots$

Explain to students the difference between 0, numbers less than 0 (negative numbers) and numbers less than 1 but still greater than 0 (such as $\frac{1}{8}$). Tell students that only subtraction patterns such as

12, 8, 4, 0, −4, −8, . . .

eventually yield 0 or a negative number.

Lesson 1.2

COMMON ERROR Students may confuse an expression with an equation. Point out that an expression has no equals sign. It is similar to an English phrase, which has no verb. Later, students will learn that an equation, which *contains* an equals sign, corresponds to an English sentence, which *has* a verb.

TEACHING TIP Stress the importance of no longer using an × to show multiplication, as it can now be confused with the variable *x*.

TEACHING TIP When writing expressions or equations for word problems, some students will benefit by using the first (unambiguous) letters of words for their variables, such as *g* for goals and *a* for assists in Example 3.

Lesson 1.3

COMMON ERROR Many students are inclined to multiply a base by its exponent. Having students write the power as a repeated multiplication as an intermediate step when evaluating should help prevent this error.

TEACHING TIP Stress the vocabulary in this lesson. Be sure that students clearly understand the terms base, exponent, and power. The term nonzero may also be new to students and may require some explanation.

Lesson 1.4

TEACHING TIP Be sure students understand the reason for the order of operations is so that everyone gets the same results to the same exercise.

COMMON ERROR Mnemonic devices such as PEMDAS work, but only as long as students use them correctly. Students must understand that multiplication and division have the *same* rank regardless of the fact that M comes before D in PEMDAS. This is similarly true for addition and subtraction.

TEACHING TIP In Example 4b, grouping symbols are implied by the fraction bar. Be sure students understand that this expression is equivalent to $(13 + 7) \div (2 \cdot 5)$.

Technology Activity 1.4

COMMON ERROR Calculators that follow the order of operations normally contain keys with parentheses on them. Some students expect that all calculators follow order of operations. These students should be taught how to distinguish which calculators follow the order of operations and which do not.

Tips for New Teachers

For use with Chapter 1

Lesson 1.5

TEACHING TIP Remind students that we solve an equation but that we simplify an expression. An equation contains an equals sign, similar to a sentence containing a verb.

TEACHING TIP Remind students to include labels with their answers, when appropriate. Point out that a label for a rate includes two units, such as mi/h.

Lesson 1.6

TEACHING TIP When finding the perimeter and the area of a figure, it becomes extremely important to label answers correctly. Remind students that perimeter is measured in *linear* units, such as inches (in.) or meters (m), and that area is measured in *square* units, such as square inches (in.^{2}) or square meters (m^2).

COMMON ERROR When finding the perimeter of a rectangle, some students may add only one length and one width. You might have those students label all four sides of the rectangle when finding perimeter and only two sides when finding area.

Problem Solving Strategies 1.7

INCLUSION You may wish to co-teach this lesson with a Special Educator who will be able to address the needs of students with language-based disabilities. These students then will have better access to solving these problems which involve a considerable amount of reading.

Lesson 1.7

TEACHING TIP Continue to stress the importance of checking all answers to problems for their reasonableness. Ask students to explain why they know their answers are correct.

CHAPTER 1

Name ______________________ Date __________

Parents as Partners

For use with Chapter 1

Chapter Overview **One way you can help your student succeed in Chapter 1 is by discussing the lesson goals in the chart below. When a lesson is completed, ask your student the following questions. "What were the goals of the lesson? What new words and formulas did you learn? How can you apply the ideas of the lesson to your life?"**

Lesson Title	*Lesson Goals*	*Key Applications*
1.1: Describing Patterns	Describe patterns using whole number operations.	• Summer Movie Club • Hawaiian Leis • Basketball • The Moon
1.2: Variables and Expressions	Evaluate variable expressions.	• Ice Hockey • Dance • Bamboo
1.3: Powers and Exponents	Write repeated multiplication using exponents. Evaluate powers.	• Biology • Dog Pedigrees • Green Goo
1.4: Order of Operations	Evaluate expressions involving two or more operations.	• Music • Video Games • Weightlifting
1.5: Equations and Mental Math	Use mental math to solve an equation.	• Homing Pigeons • Log Flumes • Cheetahs
1.6: Perimeter and Area	Use formulas to find perimeter and area.	• Wheelchair Slalom • Gardening • Globe Theatre
1.7: A Problem Solving Plan	Use a 4-step plan to solve many kinds of problems.	• Marching Band • Lasagna • Origami

Know How to Take Notes

Keeping a Notebook is the strategy featured in Chapter 1 (see page 4). Encourage your student to keep a detailed and accurate notebook. Your student may want to put in his/her notebook the following useful items: vocabulary, rules and properties, worked-out examples, symbols, formulas, diagrams, and homework assignments. Icons have been placed in your student's textbook to help identify useful notebook items. Your student should personalize his/her notebook by also including any additional items that he/she feels are necessary for better comprehension of the material.

Name ______________________________ Date __________

Parents as Partners

For use with Chapter 1

Key Ideas Your student can demonstrate understanding of key concepts by working through the following exercises with you.

Lesson	*Exercise*
1.1	The Summer Olympics are held every 4 years. Previous Olympic dates include 1988, 1992, 1996… If this pattern continues, how many Olympic Games can be held in the 21st century?
1.2	It costs \$30 for your dad to take you and three friends to a play at the theater. Use the expression $t \div p$, where t is the total amount spent and p is the number of people who attended the play to determine the cost for each ticket.
1.3	There are 2^8 seats available in each of the nine theaters at the cinema complex. How many seats are in each theater? Bonus: How many total seats are there?
1.4	Your mother and her friend take you and four friends to the movies. It costs \$7 for an adult ticket and \$4 for a student ticket. Use the expression $7a + 4s$, where a is the number of adults and s is the number of students who attend the movie, to find the total cost.
1.5	The middle school cross-country team ran 6 miles in 48 minutes. On average, how long did it take them to run each mile?
1.6	The dimensions of a tennis court are: length = 78 feet; width = 36 feet. Find the perimeter. The dimensions of a table tennis table are: length = 9 feet; width = 5 feet. Find the area.
1.7	Use the 4-step problem solving plan to solve the following problem. You are bringing some zucchini bread to the family picnic. One loaf of bread can easily be divided into 12 slices. There will be 31 people attending the picnic. How many loaves of bread do you need to bring?

Home Involvement Activity

Directions: Find the area of a surface at your home (i.e.: wall, floor, porch). Next decide how many gallons of paint will be needed to paint it, depending upon what type of surface you chose. (Porous concrete: 50–125 ft^2 per gallon; Precast concrete and brick: 150–175 ft^2 per gallon; Weathered wood: 200 ft^2 per gallon; Sealed or previously painted surface: 350–400 ft^2 per gallon.)

Answers

1.1: 25 Olympic Games **1.2:** \$6 per ticket (for 5 tickets)
1.3: 256 seats per theater; 2304 seats at the complex **1.4:** \$34 **1.5:** 8 min
1.6: tennis court: perimeter = 228 feet; table tennis table: area = 45 square feet **1.7:** 3 loaves

LESSON 1.1

Teacher's Name ________________ Class __________ Room _________ Date __________

Lesson Plan

1-day lesson (See *Pacing and Assignment Guide*, TE page 2A)

For use with pages 5–8

GOAL **Describe patterns using whole number operations.**

State/Local Objectives ____

____

✓ Check the items you wish to use for this lesson.

STARTING OPTIONS

____ Warm-Up: Transparencies

TEACHING OPTIONS

____ Notetaking Guide
____ Examples: 1–4, SE pages 5–6
____ Extra Examples: TE page 6
____ Your Turn Now Exercises: 1–4, SE page 6
____ Activity Master: CRB page 7
____ Concept Check: TE page 6
____ Getting Ready to Practice Exercises: 1–7, SE page 7

APPLY/HOMEWORK

Homework Assignment

____ Basic: Day 1: pp. 7–8 Exs. 8–13, 18–21, 23, 24, 26, 34–47
____ Average: Day 1: pp. 7–8 Exs. 12–29, 34–47
____ Advanced: Day 1: pp. 7–8 Exs. 14–18, 21–33*, 36, 37, 40, 41, 46, 47

Reteaching the Lesson

____ Practice: CRB pages 8–10 (Level A, Level B, Level C); Practice Workbook
____ Study Guide: CRB pages 11–12; Spanish Study Guide

Extending the Lesson

____ Challenge: SE page 8; CRB page 13

ASSESSMENT OPTIONS

____ Daily Quiz (1.1): TE page 8 or Transparencies
____ Test-Taking Practice: SE page 8

Notes

____

____

____

____

Lesson 1.1

LESSON 1.1

Teacher's Name ________________ Class __________ Room _________ Date __________

Lesson Plan for Block Scheduling

Half-block lesson (See *Pacing and Assignment Guide*, TE page 2A)

For use with pages 5–8

GOAL **Describe patterns using whole number operations.**

State/Local Objectives ____

____

✓ **Check the items you wish to use for this lesson.**

Chapter Pacing Guide	
Day	**Lesson**
1	**1.1;** 1.2 (begin)
2	1.2 (end); 1.3
3	1.4
4	1.5; 1.6
5	1.7
6	Ch. 1 Review and Projects

STARTING OPTIONS

____ Warm-Up: Transparencies

TEACHING OPTIONS

____ Notetaking Guide

____ Examples: 1–4, SE pages 5–6

____ Extra Examples: TE page 6

____ Your Turn Now Exercises: 1–4, SE page 6

____ Activity Master: CRB page 7

____ Concept Check: TE page 6

____ Getting Ready to Practice Exercises: 1–7, SE page 7

APPLY/HOMEWORK

Homework Assignment

____ Block Schedule: pp. 7–8 Exs. 12–29, 34–47 (with 1.2)

Reteaching the Lesson

____ Practice: CRB pages 8–10 (Level A, Level B, Level C); Practice Workbook

____ Study Guide: CRB pages 11–12; Spanish Study Guide

Extending the Lesson

____ Challenge: SE page 8; CRB page 13

ASSESSMENT OPTIONS

____ Daily Quiz (1.1): TE page 8 or Transparencies

____ Test-Taking Practice: SE page 8

Notes

____

____

____

____

____

LESSON 1.1

Name ______________________________ Date __________

Activity Master

For use before Lesson 1.1

Goal
Recognize relationships within a pattern. Then extend the pattern.

Materials
- colored pencils
- paper

Exploring Patterns

In this activity, you will explore various patterns by sketching figures and symbols.

EXPLORE **Draw the next figure in the pattern.**

1 Draw the three figures shown at the right on a piece of paper.

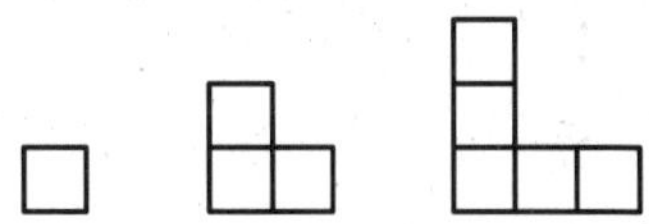

Assume the figures continue in this manner. What would you expect the next figure to look like?

2 Each figure increases in height and in length by one square as the pattern continues. The third figure has a total of 5 squares arranged in the shape of an "L". So, the fourth figure should have a total of 7 squares arranged in the shape of an "L".

Sketch the next figure.

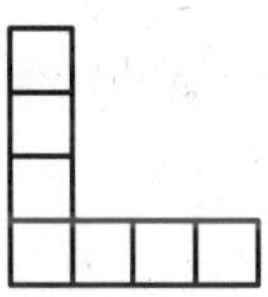

your turn now **In Exercises 1 and 2, refer to the activity above.**

1. Do you see a relationship between the first and second figure, between the second and third figure, and so on? If so, describe the relationship.

2. What would you expect the 7th and 8th figures to look like? Sketch your prediction.

3. Sketch the four symbols shown below. Assume the symbols continue in the same manner. What would you expect the next three symbols to look like? Sketch your prediction.

4. Create your own pattern. Draw five figures that form a pattern. Then have a partner sketch the next three figures. Can your partner sketch more than one pattern? Explain.

LESSON 1.1

Name ______________________________ Date ________

Practice A

For use with pages 5-8

Find the sum, difference, product, or quotient.

1. 826 + 37 **2.** 527 − 186 **3.** 142 × 28 **4.** 275 ÷ 5

Match the pattern with its description.

5. 3125, 625, 125, 25, . . . **A.** Multiply the previous number by 3.

6. 65, 55, 45, 35, . . . **B.** Subtract 10 from the previous number.

7. 1, 3, 9, 27, . . . **C.** Add 4 to the previous number.

8. 1, 5, 9, 13, . . . **D.** Divide the previous number by 5.

Describe the pattern. Then write the next three numbers.

9. 1, 3, 5, 7, . . . **10.** 44, 40, 36, 32, . . . **11.** 1, 8, 64, 512, . . .

Describe the pattern. Then draw the next figure.

12. 

13. 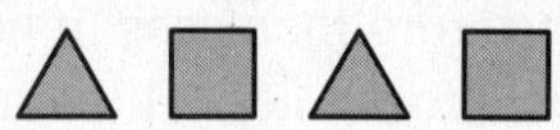

14. The owner's manual for your car states that you should change the oil every 5000 miles. If you started driving the car at 0 miles, when will the first five oil changes occur? Describe the pattern for the oil changes.

Describe the pattern. Then write the next three numbers.

15. 3, 10, 17, 24, . . . **16.** 729, 243, 81, 27, . . . **17.** 29, 26, 23, 20, . . .

Describe the pattern. Then draw the next figure.

18. 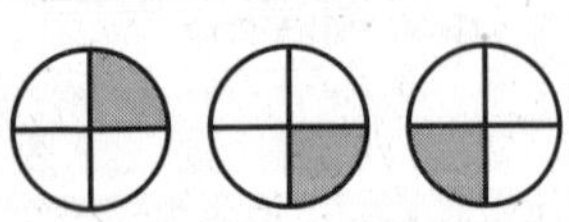

19. 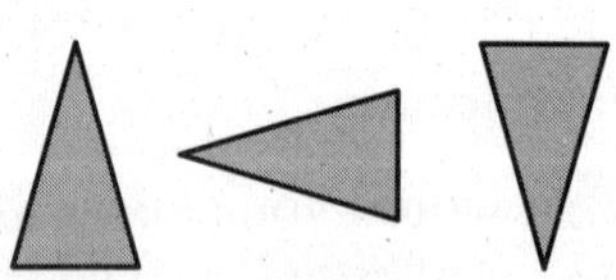

20. You are putting in a tile border on the floor around the edge of a room. The tile you have put down so far is shown below. What are the next three columns of tile that you will need to put down?

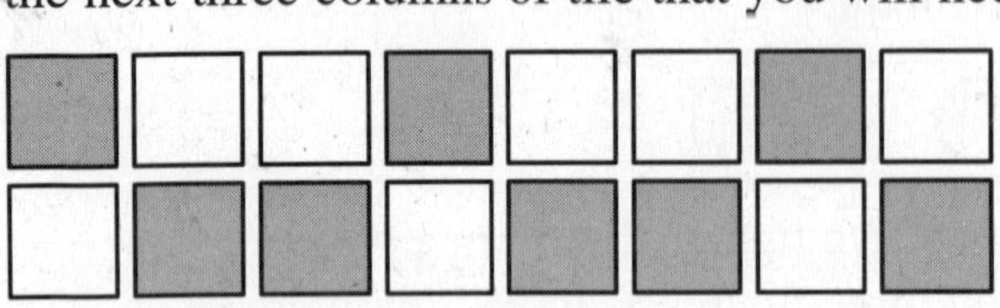

21. You decide to exercise by running three times a week. The first week you run for 15 minutes. Every two weeks you increase your running time by 5 minutes so during the third week you are running for 20 minutes. How long are you running during the ninth week?

LESSON 1.1

Name ______________________ Date __________

Practice B

For use with pages 5–8

Describe the pattern. Then write the next three numbers.

1. 1, 3, 9, 27, . . .

2. 4, 10, 16, 22, . . .

3. 21, 19, 17, 15, . . .

4. 3, 12, 48, 192, . . .

5. 6400, 3200, 1600, 800, . . .

6. 4, 20, 100, 500, . . .

7. Every four years, the month of February has 29 days instead of 28 days. The last leap year occurred in 2000. When will the next four leap years occur?

Describe the pattern. Then draw the next figure.

8.

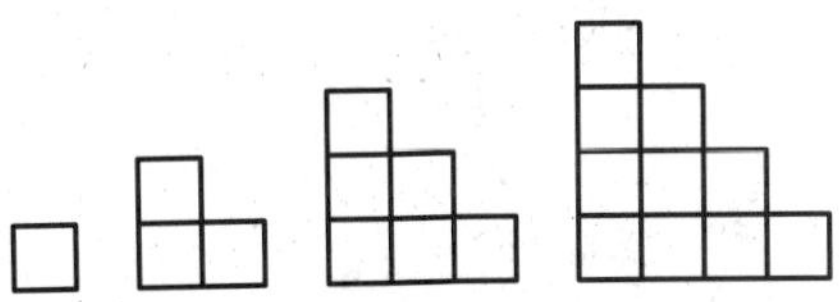

9.

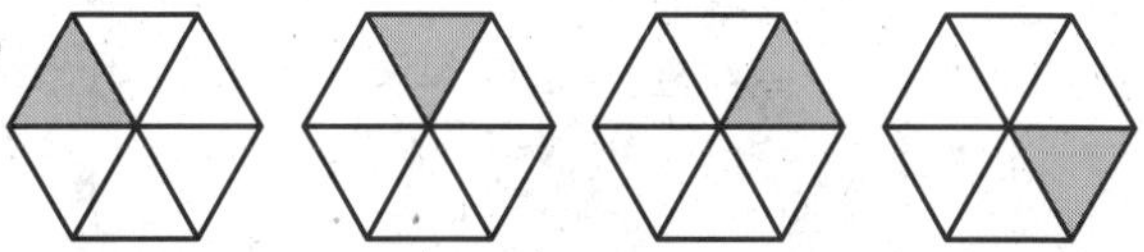

10. The owner's manual for your car states that you should have the oil changed every 3750 miles. If you started driving the car at 0 miles, when will the first five oil changes occur? Describe the pattern for the oil changes.

Describe the pattern. Then write the next three letters.

11. A, D, G, J, . . .

12. Z, X, V, T, . . .

13. A, E, I, M, . . .

Describe the pattern. Then draw the next figure.

14.

15.

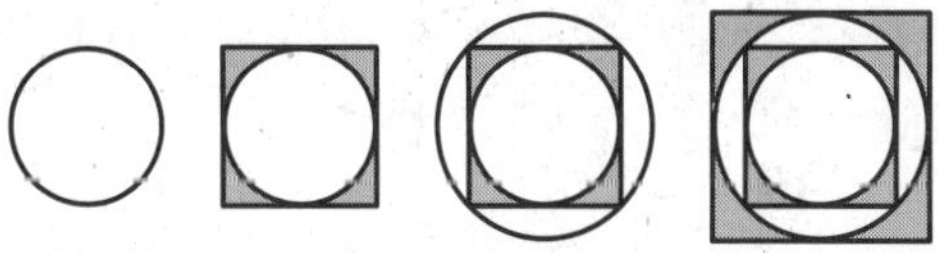

16. In the Chinese calendar, each year is named as an animal. For example, 1978 was the year of the horse. During a 12-year period, each year is named with a different animal (rat, ox, tiger, rabbit, dragon, snake, horse, goat, monkey, rooster, dog, and pig). Then the animals are used over again for the next 12 years in the same order. After 1978, what are the next five years of the horse? Describe the pattern you used to answer the question.

17. A sweater is decorated with the Southwestern-inspired design shown below. Draw the next three figures that will continue the pattern.

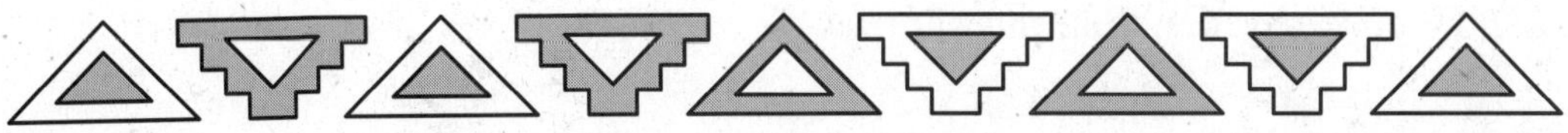

Lesson 1.1

Name ______________________________ Date __________

Practice C

For use with pages 5–8

Describe the pattern. Then write the next three numbers.

1. 1, 4, 16, 64, . . .

2. 2, 13, 24, 35, . . .

3. 74, 66, 58, 50, . . .

4. 6, 18, 54, 162, . . .

5. 1792, 896, 448, 224, . . .

6. 91, 85, 79, 73, . . .

7. Write the first four numbers in the pattern *start with 5 and repeatedly multiply by 7.*

8. You are a member of a mystery book club at a local bookstore. The club meets every other Monday. If the first meeting is on April 1, what are the dates of the next two meetings?

9. In your own words, explain how to describe and extend a numerical pattern.

Describe the pattern. Then draw the next figure.

10. 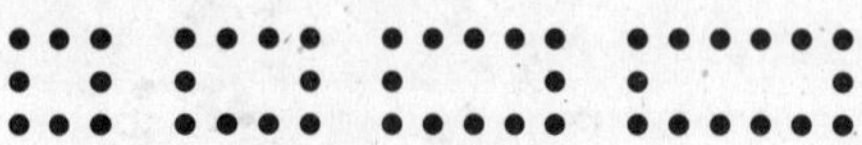

11. 

12. You are painting a border on your bathroom wall using the pattern below. Draw the next five fish to continue the pattern.

13. The beginning of a rug design is shown at the right. Complete the rug.

14. Thirteen-year cicadas are insects that appear in large groups every 13 years. Seventeen-year cicadas are insects that appear in large groups every 17 years. In 1961, a group of 13-year cicadas appeared, and in 1962, a group of 17-year cicadas appeared. Is there a year in the future when these two groups will appear at the same time? Explain how you arrived at your answer.

LESSON 1.1

Name ______________________ Date ________

Study Guide

For use with pages 5–8

GOAL **Describe patterns using whole number operations.**

EXAMPLE 1 Recognizing and Extending a Pattern

You start a summer job that pays $7 an hour. For each year worked, you will receive a $.50 an hour raise. If you keep the job for 5 years, how much will you earn per hour in the fifth year?

Solution

Pay in 1st year: $7.00

Pay in 2nd year: $7.00 + $.50 = $7.50

Pay in 3rd year: $7.50 + $.50 = $8.00

Pay in 4th year: $8.00 + $.50 = $8.50

Pay in 5th year: $8.50 + $.50 = $9.00

Answer: You will earn $9 an hour in your fifth year.

Exercises for Example 1

1. Write the first five numbers in the pattern *start with 28 and repeatedly add 5.*

2. While waiting at the bus stop, you notice that the traffic light turned red at 8:19, 8:22, and 8:25. If you assume the pattern continues, when will the light next turn red?

EXAMPLE 2 Extending a Numerical Pattern

Describe the following pattern: 36, 33, 30, 27, ... Then write the next three numbers.

You *subtract* 3 from the previous number to get the next number in the pattern.

− 3 − 3 − 3

36, 33, 30, 27, 24, 21, 18

− 3 − 3 − 3

Exercises for Example 2

Describe the pattern. Then write the next three numbers.

3. 14, 22, 30, 38, . . .

4. 67, 58, 49, 40, . . .

Name ______________________________ Date __________

Study Guide

For use with pages 5–8

EXAMPLE 3 Extending a Numerical Pattern

Describe the following pattern: 576, 288, 144, 72, ... Then write the next three numbers.

You *divide* the previous number by 2 to get the next number in the pattern.

÷ 2 ÷ 2 ÷ 2

576, 288, 144, 72, 36, 18, 9

÷ 2 ÷ 2 ÷ 2

Exercises for Example 3

Describe the pattern. Then write the next three numbers.

5. 4, 16, 64, 256, . . .

6. 729, 243, 81, 27, . . .

EXAMPLE 4 Extending a Visual Pattern

Sketch the next figure in the pattern.

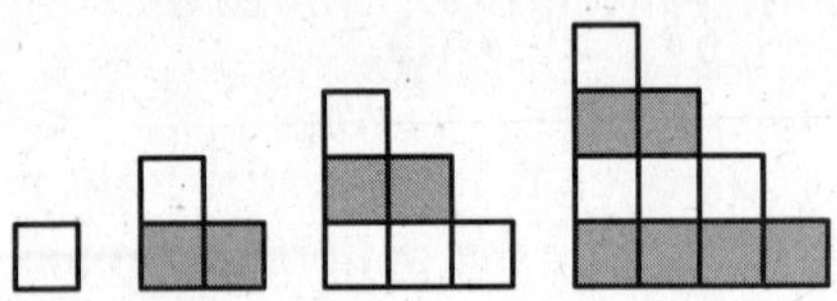

Solution

Each figure in the pattern looks like the previous figure with another row of squares added to the bottom. Also, every other row is shaded.

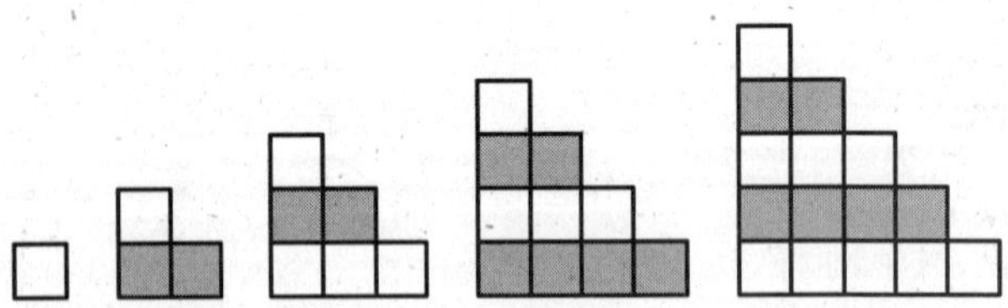

Exercises for Example 4

Sketch the next figure in the pattern.

7.

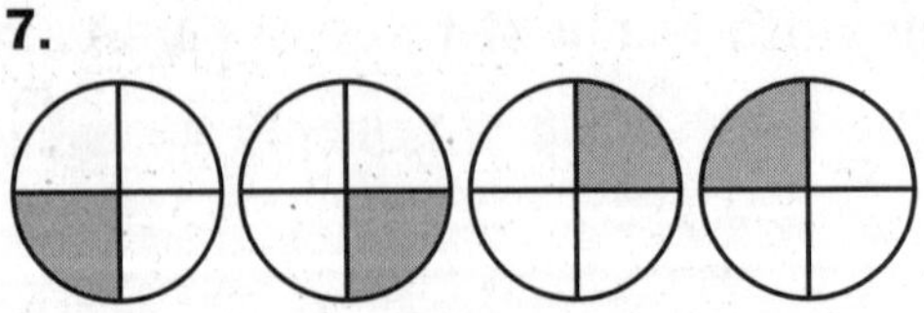

8.

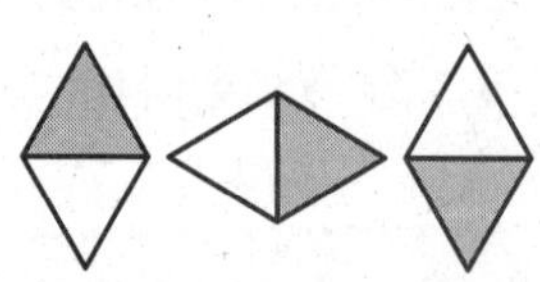

LESSON 1.1

Name ______________________________ Date __________

Challenge Practice

For use with pages 5–8

Describe the pattern. Then write the next three numbers.

1. 13, 78, 468, 2808, . . .

2. 789, 678, 567, 456, 345, . . .

3. 4, 2, 1, $\frac{1}{2}$, $\frac{1}{4}$, . . .

4. 1, 11, 7, 17, 13, 23, . . .

5. 7, 1, 8, 9, 17, 26, . . .

6. 68, 34, 17, 22, 11, 16, 8, . . .

The formulas for the area and perimeter of a square are *Area* = s^2 and *Perimeter* = 4*s*, where *s* is the length of the side of the square.

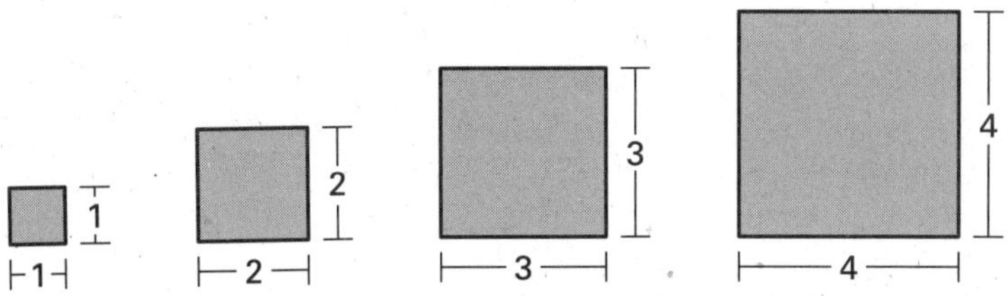

7. Calculate the perimeter of each square. Describe the pattern. Then write the next three numbers.

8. Calculate the area of each square. Describe the pattern. Then write the next three numbers.

LESSON 1.2

Teacher's Name ________________ Class __________ Room __________ Date __________

Lesson Plan

2-day lesson (See *Pacing and Assignment Guide*, TE page 2A)

For use with pages 9–12

GOAL **Evaluate variable expressions.**

State/Local Objectives ____

____

✓ **Check the items you wish to use for this lesson.**

STARTING OPTIONS

____ Homework Check (1.1): TE page 7; Answer Transparencies

____ Homework Quiz (1.1): TE page 8; Transparencies

____ Warm-Up: Transparencies

TEACHING OPTIONS

____ Notetaking Guide

____ Activity: SE page 9

____ Examples: Day 1: 1–2, SE pages 9–10; Day 2: 3, SE page 10

____ Extra Examples: TE page 10

____ Your Turn Now Exercises: Day 1: 1–8, SE pages 9–10; Day 2: none

____ Technology Activity: CRB page 16

____ Concept Check: TE page 10

____ Getting Ready to Practice Exercises: Day 1: 1–10, SE page 11; Day 2: 11, SE page 11

APPLY/HOMEWORK

Homework Assignment

____ Basic: Day 1: SRH p. 684 Exs. 5–12; pp. 11–12 Exs. 13–23;
Day 2: pp. 11–12 Exs. 12, 26–30, 41–53

____ Average: Day 1: pp. 11–12 Exs. 12, 13, 17–25, 41–44; Day 2: p. 12 Exs. 26–34, 36, 45–53

____ Advanced: Day 1: pp. 11–12 Exs. 12, 13, 19–25, 34, 37*, 38*, 41–44;
Day 2: p. 12 Exs. 28–33, 35, 36, 39*, 40*, 45–47, 52, 53

Reteaching the Lesson

____ Practice: CRB pages 17–19 (Level A, Level B, Level C); Practice Workbook

____ Study Guide: CRB pages 20–21; Spanish Study Guide

Extending the Lesson

____ Real-World Problem Solving: CRB page 22

____ Challenge: SE page 12; CRB page 23

ASSESSMENT OPTIONS

____ Daily Quiz (1.2): TE page 12 or Transparencies

____ Test-Taking Practice: SE page 12

Notes ____

LESSON 1.2

Teacher's Name ____________ Class ________ Room ________ Date ________

Lesson Plan for Block Scheduling

1-block lesson (See *Pacing and Assignment Guide*, TE page 2A)

For use with pages 9–12

GOAL **Evaluate variable expressions.**

State/Local Objectives ________________________________

____

✓ Check the items you wish to use for this lesson.

Chapter Pacing Guide	
Day	**Lesson**
1	1.1; **1.2 (begin)**
2	**1.2 (end);** 1.3
3	1.4
4	1.5; 1.6
5	1.7
6	Ch. 1 Review and Projects

STARTING OPTIONS

____ Homework Check (1.1): TE page 7; Answer Transparencies
____ Homework Quiz (1.1): TE page 8; Transparencies
____ Warm-Up: Transparencies

TEACHING OPTIONS

____ Notetaking Guide
____ Activity: SE page 9
____ Examples: Day 1: 1–2, SE pages 9–10; Day 2: 3, SE page 10
____ Extra Examples: TE page 10
____ Your Turn Now Exercises: Day 1: 1–8, SE pages 9–10; Day 2: none
____ Technology Activity: CRB page 16
____ Concept Check: TE page 10
____ Getting Ready to Practice Exercises: Day 1: 1–10, SE page 11; Day 2: 11, SE page 11

APPLY/HOMEWORK

Homework Assignment

____ Block Schedule: Day 1: pp. 11–12 Exs. 12, 13, 17–25, 41–44 (with 1.1)
Day 2: p. 12 Exs. 26–34, 36, 45–53 (with 1.3)

Reteaching the Lesson

____ Practice: CRB pages 17–19 (Level A, Level B, Level C); Practice Workbook
____ Study Guide: CRB pages 20–21; Spanish Study Guide

Extending the Lesson

____ Real-World Problem Solving: CRB page 22
____ Challenge: SE page 12; CRB page 23

ASSESSMENT OPTIONS

____ Daily Quiz (1.2): TE page 12 or Transparencies
____ Test-Taking Practice: SE page 12

Notes

____

____

Lesson 1.2

LESSON 1.2

Name ______________________________ Date __________

Technology Activity

For use with pages 9–12

GOAL Use a calculator to evaluate expressions with variables.

EXAMPLE You are making bookmarks. To make one bookmark, you use 13 inches of ribbon. The expression $13n$, where n is the number of bookmarks you make, can be used to find the total number of inches of ribbon used to make n bookmarks. If you want to make 20 bookmarks, how many inches of ribbon will you need?

Solution

You can evaluate the expression $13n$ when $n = 20$ as shown below.

Keystrokes	Display
13 [×] 20 [=]	260

Answer: If you want to make 20 bookmarks, you will need 260 inches of ribbon.

Lesson 1.2

your turn now **Use a calculator to evaluate the expression when $x = 9$ and $y = 5$.**

1. $x + 74$

2. $93 - y$

3. $14x$

4. $22y$

5. $7x + 13y$

6. $9x + 15$

7. $11x - 19y$

8. $18x - 30y$

9. $4xy$

10. $7xy$

11. You have 400 inches of ribbon. You want to make 30 bookmarks. Do you have enough ribbon? Use the expression given in the Example.

12. You have 500 inches of ribbon. You want to make 40 bookmarks. Do you have enough ribbon? Use the expression given in the Example.

13. To make one bookmark, you use 12 inches of ribbon. The expression $\frac{r}{12}$, where r is the number of inches of ribbon you have, can be used to find the total number of bookmarks you can make. You have 156 inches of ribbon. How many bookmarks can you make?

14. To make one bookmark, you use 11 inches of ribbon. The expression $\frac{r}{11}$, where r is the number of inches of ribbon you have, can be used to find the total number of bookmarks you can make. You have 231 inches of ribbon. How many bookmarks can you make?

LESSON 1.2

Name ______________________________ Date __________

Practice A

For use with pages 9–12

Complete the statement.

1. ___?___ $+ 5 = 8$

2. ___?___ $\div 4 = 6$

3. $5 \times$ ___?___ $= 60$

4. $30 \div$ ___?___ $= 2$

5. $7 +$ ___?___ $= 13$

6. $12 -$ ___?___ $= 3$

7. You ___?___ a variable expression by substituting values for the variables and then simplifying the resulting numerical expression.

Evaluate the expression when $a = 4$ and $b = 7$.

8. $a + 2$

9. $b - 1$

10. $4a$

11. $\frac{a}{2}$

12. $5 - a$

13. $\frac{21}{b}$

14. You take a guitar lesson every week. The expression $4g$, where g is the cost of one lesson, can be used to find the total cost of your lessons for the month. Find the total cost for the month if one lesson is $15.

Evaluate the expression for the given value of the variable.

15. $3 + m$ when $m = 18$

16. $\frac{20}{x}$ when $x = 4$

17. $7z$ when $z = 12$

18. $a + 11$ when $a = 29$

19. $38 - t$ when $t = 12$

20. $17y$ when $y = 8$

21. For your birthday, a friend gave you a $25 gift certificate to your favorite clothing store. The expression $p - 25$, where p is the price of an item, can be used to find the amount of money you'll have to pay in cash when you use the gift certificate. If you find a pair of jeans at the store for $42, how much do you have to pay for the jeans in cash if you use the gift certificate?

22. Your relative's house is 110 miles away. The expression $\frac{110}{r}$, where r is your driving speed, can be used to find how long it will take you to drive to your relative's house. If your driving speed is 55 miles per hour, how long will it take you to get to your relative's house?

Lesson 1.2

LESSON 1.2

Name ______________________________ Date __________

Practice B

For use with pages 9–12

Evaluate the expression for the given value of the variable.

1. $p + 13$ when $p = 8$ **2.** $23 - x$ when $x = 6$ **3.** $13y$ when $y = 7$

4. $\frac{b}{5}$ when $b = 75$ **5.** $m - 4$ when $m = 33$ **6.** $\frac{81}{a}$ when $a = 3$

7. Your part-time summer job pays $7 per hour. To find the amount of money you can earn in one week, you can evaluate the expression $7h$, where h is the number of hours you worked during the week. If you work 15 hours your first week, how much money do you earn for the week?

8. To find the number of calories from protein in a serving of food, you can evaluate the expression $4g$, where g is the number of grams of protein in a serving. If a serving of nuts contains 12 grams of protein, how many calories from protein are in the serving?

Lesson 1.2

Evaluate the expression when $a = 3$, $b = 8$, $m = 24$, and $q = 35$.

9. $b + q$ **10.** am **11.** $q - m$

12. $\frac{m}{b}$ **13.** $m - b$ **14.** $a + q$

15. During the winter, weather forecasters give the actual outdoor temperature and the "wind chill" temperature. This is because the blowing wind can make it feel colder than it actually is. The expression $a - w$, where w is the wind chill temperature and a is the actual temperature, can be used to find how many degrees colder it feels outside when the wind is blowing. If the actual temperature is 35° F and the wind chill temperature is 16° F, how many degrees colder does it feel because of the wind?

16. The Pennsylvania Turnpike is a 528-mile long roadway that spans the entire state. To find how long it will take you to drive across the state on the turnpike, you can evaluate the expression $\frac{L}{r}$, where L is the length of the Pennsylvania Turnpike and r is your driving speed. If you maintain a driving speed of 66 miles per hour, how many hours will it take you to drive across the state?

17. Which two expressions are the same?

A. $\frac{a}{b}$ **B.** $a - b$

C. $a \div b$ **D.** $\frac{b}{a}$

LESSON 1.2

Name ______________________________ Date __________

Practice C

For use with pages 9–12

Evaluate the expression when $x = 4$, $y = 11$, $a = 36$, and $b = 42$.

1. $a + b$

2. $b - x$

3. by

4. $\frac{a}{x}$

5. ax

6. $b - a$

7. You take your car to your mechanic to have it repaired. The expression $p + \ell$, where p is the cost of parts and ℓ is the cost of labor, can be used to find the amount you have to pay for the repairs. If the parts cost \$125 and the labor costs \$36, how much do you have to pay for repairs?

8. Find the value of x that makes the expression $x - 11$ equal to 45.

9. Your favorite band is playing a concert in your hometown. You purchase your ticket from a local ticket outlet, which charges a service fee for processing your order. The expression $c - p$, where c is the amount charged by the outlet and p is the ticket price, can be used to find the amount of the service fee. If the outlet charges \$42 and the ticket price is \$35, what is the amount of the service fee?

Evaluate the expression when $m = 14$, $n = 15$, $s = 41$, and $t = 70$.

10. $n + s$

11. $t - s$

12. nt

13. $\frac{t}{m}$

14. mn

15. $s - m$

16. A pizza shop offers a lunch buffet from 11 A.M. to 2 P.M. To find the total cost of lunch for a group of people, you can use the expression bn, where b is the price of the buffet for one person and n is the number of people eating at the buffet. If you have 11 people in your group all ordering the buffet and the buffet is \$7 per person, what is the total cost of lunch for the group?

17. Explain why you should not use the symbol $\times$ in an expression that contains variables.

Evaluate the expression when $a = 5$, $b = 4$, and $c = 12$.

18. $7ab$

19. abc

20. $a + b + c$

21. An antelope can run 16 miles per hour faster than an elk. If an antelope runs at a speed of a miles per hour, which expression can you use to find the speed of an elk?

A. $\frac{a}{16}$

B. $16a$

C. $a + 16$

D. $a - 16$

LESSON 1.2

Name ______________________ Date ________

Study Guide

For use with pages 9–12

GOAL Evaluate variable expressions.

VOCABULARY

A **variable** is a letter that is used to represent one or more numbers.
A **variable expression** consists of numbers, variables, and operations.
To **evaluate** a variable expression, substitute values for the variables and then simplify the resulting numerical expression.

EXAMPLE 1 Evaluating Variable Expressions

a. Evaluate $16 + a$ when $a = 6$. **b.** Evaluate $11 - b$ when $b = 4$.

Solution

a. $16 + a = 16 + \mathbf{6}$ Substitute 6 for *a*.
$= 22$ Add.

b. $11 - b = 11 - \mathbf{4}$ Substitute 4 for *b*.
$= 7$ Subtract.

Exercises for Example 1

Evaluate the expression when $x = 5$ and $y = 10$.

1. $x + 7$ **2.** $y - 3$ **3.** $13 + y$

4. $18 - x$ **5.** $y + 10$ **6.** $x - 4$

EXAMPLE 2 Evaluating Variable Expressions

a. Evaluate $6p$ when $p = 7$. **b.** Evaluate $\frac{72}{g}$ when $g = 8$.

Solution

a. $6p = 6(\mathbf{7})$ Substitute 7 for *p*.
$= 42$ Multiply.

b. $\frac{72}{g} = \frac{72}{\mathbf{8}}$ Substitute 8 for *g*.
$= 9$ Divide.

Lesson 1.2

Name ______________________ Date __________

Study Guide

For use with pages 9–12

Exercises for Example 2

Evaluate the expression when $w = 4$ and $m = 6$.

7. $10w$ **8.** $\frac{72}{m}$ **9.** $8m$

10. $\frac{28}{w}$ **11.** $\frac{m}{3}$ **12.** $15w$

EXAMPLE 3 Evaluating Variable Expressions with Two Variables

You go to the store to buy notebooks for school. To find the total cost of the notebooks, you can evaluate the expression np, where n is the number of notebooks and p is the price of one notebook. Find the total cost of 5 notebooks that cost \$3 per notebook.

Solution

$np = \mathbf{5}(\mathbf{3})$ Substitute 5 for n and 3 for p.

$= 15$ Multiply.

Answer: The total cost is \$15.

Exercises for Example 3

Evaluate the expression when $a = 3$, $b = 24$, $c = 12$, and $d = 4$.

13. ad **14.** $\frac{c}{d}$ **15.** $b + d$

16. $b - c$ **17.** $\frac{b}{a}$ **18.** $c - a$

LESSON 1.2

Name ______________________________ Date ________

Real-World Problem Solving

For use with pages 9–12

Teaching

The job of a teacher involves many tasks. Some of the tasks include preparing lessons for each class, teaching classes, and grading papers in a timely manner. The number of classes a teacher has depends on the period of time a class takes and the length of a school day. Most middle school teachers have between 4 to 7 classes to teach per day.

In Exercises 1–3, use the following information.

Mr. Perez teaches math in a middle school. He has four different classes to prepare for the first day of school. He wants each student to have 1 math book, 1 workbook, and 2 folders. Mr. Perez divides the math books into four piles according to how many students there are in each class. He then divides the workbooks into four piles the same way. Next he must divide the folders. Mr Perez uses the variable expression $2n$, where n is the number of students in a class, to find the number of folders needed for each class. The class sizes are shown below.

Class 1: 32 students Class 3: 29 students

Class 2: 30 students Class 4: 33 students

1. How many folders are needed for each class?

2. Find the total number of math books, workbooks, and folders that are needed for Class 2.

3. Before Class 4 begins, Mr. Perez counts how many folders he has left. He counts 68 folders. Will there be enough folders for the students in Class 4?

LESSON 1.2

Name ______________________________ Date __________

Challenge Practice

For use with pages 9–12

Evaluate the expression when $a = 12$, $b = 4$, and $c = 6$.

1. $\frac{a + b}{c}$

2. $\frac{ac}{b}$

3. $8c - 12b$

Evaluate the expression when $x = 8$, $y = 11$, and $z = 14$.

4. $x + 2y + 5z$

5. $xz - \frac{55}{y}$

6. $\frac{120}{x} + yx - \frac{3z}{7}$

To find the number of points scored by the place kicker in a football game, you can evaluate the expression $3f + e$, where f is the number of field goals scored and e is the number of extra points.

7. Find the number of points scored by a kicker with 2 field goals and 3 extra points.

8. A kicker had 3 field goals in a game and scored a total of 13 points. How many extra points did the kicker score?

Lesson 1.2

LESSON 1.3

Teacher's Name ________________ Class __________ Room _________ Date __________

Lesson Plan

1-day lesson (See *Pacing and Assignment Guide*, TE page 2A)

For use with pages 13–17

GOAL **Write repeated multiplication using exponents.**

State/Local Objectives ____

____

✓ Check the items you wish to use for this lesson.

STARTING OPTIONS

____ Homework Check (1.2): TE page 11; Answer Transparencies

____ Homework Quiz (1.2): TE page 12; Transparencies

____ Warm-Up: Transparencies

TEACHING OPTIONS

____ Notetaking Guide

____ Hands-on Activity: SE page 13

____ Examples: 1–3, SE pages 14–15

____ Extra Examples: TE page 15

____ Your Turn Now Exercises: 1–9, SE page 15

____ Concept Check: TE page 15

____ Getting Ready to Practice Exercises: 1–11, SE page 16

APPLY/HOMEWORK

Homework Assignment

____ Basic: Day 1: pp. 16–17 Exs. 12–21, 24–29, 34, 35, 37, 38, 45, 49–57

____ Average: Day 1: pp. 16–17 Exs. 14–21, 28–35, 39–45, 49–57

____ Advanced: Day 1: pp. 16–17 Exs. 19–23, 30–33, 36–52*, 56, 57

Reteaching the Lesson

____ Practice: CRB pages 26–28 (Level A, Level B, Level C); Practice Workbook

____ Study Guide: CRB pages 29–30; Spanish Study Guide

Extending the Lesson

____ Challenge: SE page 17; CRB page 31

ASSESSMENT OPTIONS

____ Daily Quiz (1.3): TE page 17 or Transparencies

____ Test-Taking Practice: SE page 17

Notes

____

____

LESSON 1.3

Lesson Plan for Block Scheduling

Half-block lesson (See *Pacing and Assignment Guide*, TE page 2A)

For use with pages 13–17

GOAL **Write repeated multiplication using exponents.**

State/Local Objectives ____________________

____

✓ Check the items you wish to use for this lesson.

Chapter Pacing Guide	
Day	**Lesson**
1	1.1; 1.2 (begin)
2	1.2 (end); **1.3**
3	1.4
4	1.5; 1.6
5	1.7
6	Ch. 1 Review and Projects

STARTING OPTIONS

____ Homework Check (1.2): TE page 11; Answer Transparencies

____ Homework Quiz (1.2): TE page 12; Transparencies

____ Warm-Up: Transparencies

TEACHING OPTIONS

____ Notetaking Guide

____ Hands-on Activity: SE page 13

____ Examples: 1–3, SE pages 14–15

____ Extra Examples: TE page 15

____ Your Turn Now Exercises: 1–9, SE page 15

____ Concept Check: TE page 15

____ Getting Ready to Practice Exercises: 1–11, SE page 16

APPLY/HOMEWORK

Homework Assignment

____ Block Schedule: pp. 16–17 Exs. 14–21, 28–35, 39–45, 49–57 (with 1.2)

Reteaching the Lesson

____ Practice: CRB pages 26–28 (Level A, Level B, Level C); Practice Workbook

____ Study Guide: CRB pages 29–30; Spanish Study Guide

Extending the Lesson

____ Challenge: SE page 17; CRB page 31

ASSESSMENT OPTIONS

____ Daily Quiz (1.3): TE page 17 or Transparencies

____ Test-Taking Practice: SE page 17

Notes

____

____

____

____

LESSON 1.3

Name ______________________ Date ________

Practice A

For use with pages 13–17

Complete the statement.

1. $4 \cdot 4 = \underline{\ ?\ }$

2. $7 \cdot 7 = \underline{\ ?\ }$

3. $12 \cdot 12 = \underline{\ ?\ }$

4. A __?__ is a way of writing repeated multiplication.

Match the description in words with the power.

5. five to the fourth power

6. seven to the ninth power

7. nine to the seventh power

8. four to the fifth power

A. 4^5

B. 5^4

C. 7^9

D. 9^7

Write the product as a power.

9. $8 \cdot 8 \cdot 8$

10. $3 \cdot 3 \cdot 3 \cdot 3$

11. $6 \cdot 6 \cdot 6 \cdot 6 \cdot 6$

12. $y \cdot y \cdot y$

13. $m \cdot m \cdot m \cdot m \cdot m \cdot m$

14. $n \cdot n$

Evaluate the power.

15. 2^3

16. 9^2

17. 5^3

Evaluate the expression for the given value of the variable.

18. x^2 when $x = 7$

19. a^4 when $a = 3$

20. n^6 when $n = 2$

21. In the metric system of measurement, one meter is the same as 10^2 centimeters. Evaluate the power to find the number of centimeters in one meter.

22. Computers store information in units called bits and bytes. There are 8 bits in one byte of data. Write the number of bits in one byte as a power of 2.

23. An area rug is shown below. Write the number of small squares in the rug as a power. How many squares is this?

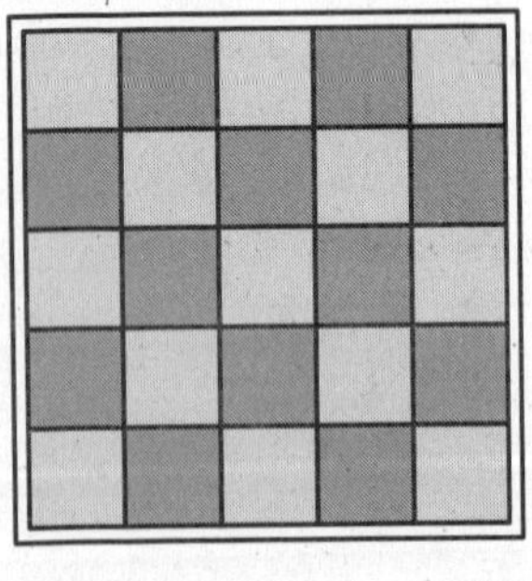

Lesson 1.3

LESSON 1.3

Name ______________________________ Date __________

Practice B

For use with pages 13–17

Write the power in words.

1. 3^8 **2.** 9^4 **3.** y^5

Write the product as a power.

4. $2 \cdot 2 \cdot 2 \cdot 2 \cdot 2 \cdot 2$ **5.** $13 \cdot 13 \cdot 13$ **6.** $6 \cdot 6 \cdot 6 \cdot 6$

7. $r \cdot r \cdot r$ **8.** $a \cdot a \cdot a \cdot a \cdot a \cdot a \cdot a$ **9.** $t \cdot t \cdot t \cdot t$

Evaluate the power.

10. 2^5 **11.** 7^4 **12.** 10^3

13. 3^5 **14.** 6^3 **15.** 1^4

Evaluate the expression for the given value of the variable.

16. y^2 when $y = 12$ **17.** b^5 when $b = 4$ **18.** w^3 when $w = 15$

19. v^6 when $v = 3$ **20.** s^4 when $s = 10$ **21.** m^8 when $m = 2$

Write the number as a power.

22. 25 **23.** 121 **24.** 64

25. Write each number in the pattern as a power: 1, 4, 9, 16, 25, . . .

26. Computers store information in units called bits and bytes. A kilobyte is 2^{10} bytes. Evaluate the power to find the number of bytes in a kilobyte.

27. In the metric system of weight, there are 1000 milligrams in one gram. Write the number of milligrams in one gram as a power.

28. The solid figure below is called a cube and is made of blocks that are all the same size. How many blocks make up the cube? Write the number of blocks that make up the cube as a power.

Lesson 1.3

LESSON 1.3

Name ______________________________ Date __________

Practice C

For use with pages 13–17

Write the power in words.

1. 7^{11} **2.** m^4 **3.** t^{12}

Write the product as a power.

4. $1 \cdot 1 \cdot 1 \cdot 1$ **5.** $4 \cdot 4 \cdot 4 \cdot 4 \cdot 4$ **6.** $t \cdot t \cdot t \cdot t \cdot t \cdot t \cdot t \cdot t \cdot t$

Evaluate the power.

7. 6^5 **8.** 2^9 **9.** 11^3

10. 0^{10} **11.** 1^{12} **12.** 5^5

Evaluate the expression for the given value of the variable.

13. a^1 when $a = 24$ **14.** t^6 when $t = 4$ **15.** n^3 when $n = 16$

16. ℓ^6 when $\ell = 5$ **17.** c^4 when $c = 20$ **18.** u^8 when $u = 1$

Write the number as a power.

19. 625 **20.** 400 **21.** 1

22. 32 **23.** 125 **24.** 27

25. Evaluate the powers 10^1, 10^2, 10^3, and 10^4. How does each exponent relate to the evaluated power? Find 10^5, 10^6, and 10^7 without performing the multiplication.

26. Computers store information in units called bits and bytes. A megabyte is 2^{20} bytes. Evaluate the power to find the number of bytes in a megabyte.

27. The solid figure below is called a cube and is made of blocks that are all the same size. How many blocks make up the cube? Write the number of blocks that make up the cube as a power.

Evaluate the expression for the given value of the variable.

28. 2^y when $y = 6$ **29.** 3^x when $x = 4$ **30.** 5^a when $a = 2$

Lesson 1.3

LESSON 1.3

Name ______________________ Date __________

Study Guide

For use with pages 13–17

GOAL **Write repeated multiplication using exponents.**

VOCABULARY

A **power** is a way of writing repeated multiplication.
The **base** of a power is the factor, and the **exponent** of a power is the number of times the factor is used.

base ⟶ exponent

$7^3 = 7 \cdot 7 \cdot 7$ 7 is a factor 3 times.

power

EXAMPLE 1 Writing Powers

Write the product as a power.

a. $3 \cdot 3 \cdot 3 \cdot 3 \cdot 3 \cdot 3$ **b.** $b \cdot b \cdot b \cdot b$

Solution

a. $3 \cdot 3 \cdot 3 \cdot 3 \cdot 3 \cdot 3 = 3^6$ 3 is a factor 6 times.

b. $b \cdot b \cdot b \cdot b = b^4$ *b* is a factor 4 times.

Exercises for Example 1

Write the product as a power.

1. $10 \cdot 10 \cdot 10$ **2.** $z \cdot z \cdot z \cdot z \cdot z \cdot z \cdot z$

3. $8 \cdot 8 \cdot 8 \cdot 8 \cdot 8$ **4.** $v \cdot v$

EXAMPLE 2 Evaluating Powers

Evaluate the power.

a. 3^3 **b.** 9^2 **c.** 2^4

Solution

a. $3^3 = 3 \cdot 3 \cdot 3$ Write 3 as a factor 3 times.
$= 27$ Multiply.

b. $9^2 = 9 \cdot 9$ Write 9 as a factor 2 times.
$= 81$ Multiply.

c. $2^4 = 2 \cdot 2 \cdot 2 \cdot 2$ Write 2 as a factor 4 times.
$= 16$ Multiply.

Name ______________________________ Date __________

Study Guide

For use with pages 13–17

Exercises for Example 2

Evaluate the power.

5. 4^5 **6.** 8^3 **7.** 5^1

EXAMPLE 3 Evaluating Powers with Variables

a. Evaluate y^3 when $y = 2$. **b.** Evaluate k^5 when $k = 3$.

Solution

a. $y^3 = \mathbf{2}^3$ Substitute 2 for y.

$= 2 \cdot 2 \cdot 2$ Write 2 as a factor 3 times.

$= 8$ Multiply.

b. $k^5 = \mathbf{3}^5$ Substitute 3 for k.

$= 3 \cdot 3 \cdot 3 \cdot 3 \cdot 3$ Write 3 as a factor 5 times.

$= 243$ Multiply.

Exercises for Example 3

8. Evaluate t^{10} when $t = 2$. **9.** Evaluate r^5 when $r = 6$.

LESSON 1.3

Name ______________________________ Date __________

Challenge Practice

For use with pages 13–17

Write the number or expression as a power.

1. 100,000

2. $(x + 1)(x + 1)(x + 1)(x + 1)$

Evaluate the expression.

3. $3^4 + 6^2$

4. $8^3 - 5^3$

5. $2^7 + 4^4 - 3^5$

6. $7^3 - 9^2 + 1^{12}$

Write the number as a power in two different ways.

7. 16

8. 81

9. 729

10. 1,000,000

Lesson 1.3

LESSON 1.4

Teacher's Name ________________ Class __________ Room __________ Date __________

Lesson Plan

2-day lesson (See *Pacing and Assignment Guide*, TE page 2A)

For use with pages 18–23

GOAL **Evaluate expressions involving two or more operations.**

State/Local Objectives ____

____

✓ **Check the items you wish to use for this lesson.**

STARTING OPTIONS

____ Homework Check (1.3): TE page 16; Answer Transparencies

____ Homework Quiz (1.3): TE page 17; Transparencies

____ Warm-Up: Transparencies

TEACHING OPTIONS

____ Notetaking Guide

____ Examples: Day 1: 1–3, SE pages 18–19; Day 2: 4–5, SE pages 19–20

____ Extra Examples: TE pages 19–20

____ Your Turn Now Exercises: Day 1: 1–5, SE page 19; Day 2: 6–7, SE page 19

____ Technology Activity: SE page 23

____ Keystrokes for Technology Activity 1.4 on SE page 23: CRB page 34

____ Concept Check: TE page 20

____ Getting Ready to Practice Exercises: Day 1: Exs. 1–5, 10, 12, 13, SE page 20; Day 2: Exs. 6–9, 11, 14

APPLY/HOMEWORK

Homework Assignment

____ Basic: Day 1: pp. 21–22 Exs. 15, 16, 21, 23–26, 33, 39–43
Day 2: pp. 21–22 Exs. 17–20, 22, 28–32, 44–49

____ Average: Day 1: pp. 21–22 Exs. 16, 21, 23–27, 33–35, 39–43
Day 2: pp. 21–22 Exs. 18–20, 22, 28–32, 44–49

____ Advanced: Day 1: pp. 21–22 Exs. 21, 23, 25–27, 33–41*
Day 2: pp. 21–22 Exs. 19, 20, 22, 28–32, 42, 43, 48, 49

Reteaching the Lesson

____ Practice: CRB pages 35–37 (Level A, Level B, Level C); Practice Workbook

____ Study Guide: CRB pages 38–39; Spanish Study Guide

Extending the Lesson

____ Challenge: SE page 21; CRB page 40

ASSESSMENT OPTIONS

____ Daily Quiz (1.4): TE page 22 or Transparencies

____ Test-Taking Practice: SE page 22

____ Quiz (1.1–1.4): SE page 25; Assessment Book page 7

Notes ____

LESSON 1.4

Teacher's Name ______________ Class __________ Room __________ Date __________

Lesson Plan for Block Scheduling

1-block lesson (See *Pacing and Assignment Guide,* TE page 2A)

For use with pages 18–23

GOAL **Evaluate expressions involving two or more operations.**

State/Local Objectives ____

____

✓ **Check the items you wish to use for this lesson.**

Chapter Pacing Guide	
Day	**Lesson**
1	1.1; 1.2 (begin)
2	1.2 (end); 1.3
3	**1.4**
4	1.5; 1.6
5	1.7
6	Ch. 1 Review and Projects

STARTING OPTIONS

____ Homework Check (1.3): TE page 16; Answer Transparencies

____ Homework Quiz (1.3): TE page 17; Transparencies

____ Warm-Up: Transparencies

TEACHING OPTIONS

____ Notetaking Guide

____ Examples: 1–5, SE pages 18–20

____ Extra Examples: TE pages 19–20

____ Your Turn Now Exercises: 1–7, SE page 19

____ Technology Activity: SE page 23

____ Keystrokes for Technology Activity 1.4 on SE page 23: CRB page 34

____ Concept Check: TE page 20

____ Getting Ready to Practice Exercises: 1–14, SE page 20

APPLY/HOMEWORK

Homework Assignment

____ Block Schedule: pp. 21–22 Exs. 16, 18–35, 39–49

Reteaching the Lesson

____ Practice: CRB pages 35–37 (Level A, Level B, Level C); Practice Workbook

____ Study Guide: CRB pages 38–39; Spanish Study Guide

Extending the Lesson

____ Challenge: SE page 21; CRB page 40

ASSESSMENT OPTIONS

____ Daily Quiz (1.4): TE page 22 or Transparencies

____ Test-Taking Practice: SE page 22

____ Quiz (1.1–1.4): SE page 25; Assessment Book page 7

Notes ____

____

____

LESSON 1.4

Name ______________________________ Date __________

Technology Activity Keystrokes

For use with Technology Activity 1.4 on page 23

TI-34 II

3 [( ] 7 [+] 6 [ )] [ENTER =]

7 [×] 3 [^] 3 [ENTER =]

TI-73

3 [( ] 7 [+] 6 [ )] [ENTER]

7 [×] 3 [^] 3 [ENTER]

Lesson 1.4

LESSON 1.4

Name ______________________ Date __________

Practice A

For use with pages 18–23

Evaluate the power.

1. 2^4

2. 3^3

3. 10^2

Complete the statement.

4. Parentheses, brackets, and __?__ are grouping symbols.

For each expression, identify the operation that is performed first.

5. $4 + 3 \cdot 5$

6. $\frac{9 + 7}{2}$

7. $(12 + 6) \div 6$

For each expression, identify the operation that is performed last.

8. $8(11 - 6)$

9. $3 + 4^2$

10. $\frac{10}{5 - 3}$

Evaluate the expression.

11. $8 + 7 \cdot 4$

12. $10 - 3 + 6 - 2$

13. $6 + 8 \div 2$

14. $6(14 - 1)$

15. $(6 - 2^2)(1 + 7)$

16. $\frac{18 - 6}{3}$

17. $16 + 3^3$

18. $40 - 6^2 + 7$

19. $24 \div 2^2$

Evaluate the expression when $a = 12$ and $b = 5$.

20. $15 \div b + a$

21. $a - 20 \div b$

22. $b^2 - a + 2$

23. The number system we use is called the base-ten system. One of the properties of this system is that a number can be written in an expanded notation that uses powers of ten. The number below is written in expanded notation. Evaluate the expression. What is the number?

$$5 \times 10^3 + 8 \times 10^2 + 3 \times 10^1 + 7$$

24. Your local cable television company charges a one-time $50 hookup fee and then charges $45 per month for extended basic cable service. Write and evaluate an expression to find the cost of getting cable for the first year.

25. Entrance to a popular amusement park costs $40 for a person who is 4 feet tall or taller and $20 for a person who is shorter than 4 feet. Write and evaluate an expression to determine the total cost for the following group to enter the park: 5 people who are over 4 feet tall and 3 people who are under 4 feet tall.

LESSON 1.4

Name ______________________________ Date ________

Practice B

For use with pages 18–23

Evaluate the expression.

1. $10 + 6 \cdot 8$
2. $35 - 20 \div 5$
3. $18(11 - 6)$
4. $28 \div (16 - 9)$
5. $\frac{14 + 10}{7 - 4}$
6. $33 \div (3^2 + 2)$
7. $5(7 + 5)^2$
8. $(5 - 3)^3 + 12 \div 4$
9. $11(8 + 3^2) - 15$

Evaluate the expression when $x = 4$, $y = 12$, and $z = 9$.

10. $3xy - 7$
11. $3x^2 + y$
12. $\frac{5y}{x}$
13. $4z^2 - x$
14. $\frac{y - z}{3}$
15. $(y - x)^2 + 14$
16. $\frac{x + y}{z - 5}$
17. $(x^2 - y)(z + 4)$
18. $36 \div z - x$

Match the given expression with the expression that has the same value.

19. $36 - 5^2 + 7$ A. $2^2 \cdot 8 - 10$
20. $27 - 2 \cdot 4 + 3$ B. $(4^2 + 38) \div 3$
21. $12(8 - 3)$ C. $(5 + 3)^2 - 4$

22. Your local phone company offers DSL service for connecting to the Internet. There is a one-time \$100 hookup fee and monthly charges are \$40 for the service. Write and evaluate an expression to find the cost of getting DSL service for the first year.

23. At the beginning of the workday, a cashier will start out with money in his or her cash register so that change can be made. A cash register contains 50 one-dollar bills, 30 five-dollar bills, and 8 ten-dollar bills. Write and evaluate an expression that will give you the total amount of money in the cash register.

LESSON 1.4

Name ______________________ Date __________

Practice C

For use with pages 18–23

Evaluate the expression.

1. $2 \cdot 11 \div (14 - 3)$

2. $12(9 - 7)^2$

3. $\frac{4 \cdot 9}{13 - 7}$

4. $(4 + 6)^2 \div 25$

5. $(12 + 8)(7 - 4)^3$

6. $[(7 - 3)^2 + 2] \div 3$

7. $5(18 - 3^2) + 12$

8. $32 \div (11 - 9)^3 - 1$

9. $\frac{28 - 4^2}{1 + 2}$

Evaluate the expression when $x = 5$, $y = 14$, and $z = 6$.

10. $10y - x^2$

11. $3x^2 + 2y$

12. $\frac{5z}{3x}$

13. $4z^2 - x + 3y$

14. $\frac{xz}{y - 8}$

15. $(y - x)^2 + 8z$

16. $\frac{x + y}{z + 13}$

17. $(x^2 - 25)(y + z)$

18. $(xz)^2 + y$

19. Insert grouping symbols into the expression $11 + 3 \cdot 17 - 5$ to make it equal to 47.

20. A place-kicker can score points in three different ways: making a touchdown, kicking a field goal, and kicking the extra point after a touchdown. A touchdown is worth 6 points, a field goal is worth 3 points, and an extra point is worth 1 point. If a place-kicker scored 6 touchdowns, 33 extra points, and 18 field goals in one season, how many points did the player score for the season?

21. Choose the two expressions that, when evaluated, use the same operations in the same order.

A. $(8 - 2)^2 \div (8 + 4)$

B. $\frac{34 - 2^2}{3 + 7}$

C. $(30 - 4^2) \div (5 + 2)$

Name ______________________ Date __________

Study Guide

For use with pages 18–23

GOAL **Evaluate expressions involving two or more operations.**

Order of Operations

1. Evaluate expressions inside grouping symbols.
2. Evaluate powers.
3. Multiply and divide from left to right.
4. Add and subtract from left to right.

EXAMPLE 1 Following Order of Operations

Tickets to a carnival cost \$5 each. You then pay \$2 for each of 7 rides at the carnival. Evaluate the expression $5 + 2 \times 7$ to find the total amount you spent.

Solution

$5 + 2 \times 7 = 5 + 14$ First multiply 2 and 7.

$= 19$ Then add 5 and 14.

Answer: The total amount you spent is \$19.

Exercise for Example 1

1. A gym has a \$45 membership fee and charges \$20 for each month. Write and evaluate an expression to find the cost to join for one year.

EXAMPLE 2 Evaluating a Variable Expression

Evaluate the expression $m + 5n^2$ when $m = 16$ and $n = 3$.

Solution

$m + 5n^2 = \mathbf{16} + 5(\mathbf{3}^2)$ Substitute 16 for *m* and 3 for *n*.

$= 16 + 5(9)$ Evaluate the power.

$= 16 + 45$ Multiply 5 and 9.

$= 61$ Add 16 and 45.

Exercises for Example 2

Evaluate the expression when $x = 4$, $y = 7$, $w = 2$, and $z = 10$.

2. $\frac{z - w}{x}$

3. $3x + 5w^2$

4. $21 \div y + 2z$

LESSON 1.4 Continued

Name ______________________________ Date __________

Study Guide

For use with pages 18–23

EXAMPLE 3 Using the Left-to-Right Rule

a. $21 + 9 - 15 - 4 = 30 - 15 - 4$ Add 21 and 9.

$= 15 - 4$ Subtract 15 from 30.

$= 11$ Subtract 4 from 15.

b. $3 \times 6 \times 2 = 18 \times 2$ Multiply 3 and 6.

$= 36$ Multiply 18 and 2.

Exercises for Example 3

Evaluate the expression.

5. $44 + 23 - 51 + 10$

6. $64 \div 8 \div 4$

7. $34 + 8 \div 2$

8. $9 \times 9 + 18 \div 3$

EXAMPLE 4 Using Grouping Symbols

a. $\frac{4 \cdot 9}{25 - 13} = \frac{36}{12}$ Multiply 4 and 9. Subtract 13 from 25.

$= 3$ Divide 36 by 12.

b. $(5 + 3)^2 \cdot (8 - 6) = 8^2 \cdot 2$ Add and subtract inside parentheses.

$= 64 \cdot 2$ Evaluate the power.

$= 128$ Multiply 64 and 2.

Exercises for Example 4

Evaluate the expression.

9. $(17 - 9) \times 10$

10. $\frac{37 - 19}{3 + 3}$

11. $21 \div 3 + (5^2 - 13)$

Lesson 1.4

LESSON 1.4

Name ______________________________ Date __________

Challenge Practice

For use with pages 18–23

Evaluate the expression.

1. $[8 \div (3 - 1)^2] + 10$

2. $\dfrac{4^3 - 2^3}{(4 - 2)^3}$

3. $1 + 2 \cdot 3 \cdot 4 \div 6 - 5 + 7 \cdot 8$

4. $9 \cdot 8 - 7 \cdot 6 - 5 - 4 \cdot 3 \div 2 + 1$

Insert the operation of addition, subtraction, multiplication, or division in each blank to make each equation true.

5. $(3^2$ __?__ $4)$ __?__ $5 = 1$

6. 150 __?__ $(15$ __?__ $5) = 2$

7. 15 __?__ 10 __?__ 2 __?__ $6 = 14$

8. 20 __?__ $(8$ __?__ $2)^2$ __?__ 7 __?__ $5 = 1$

Lesson 1.4

LESSON 1.5

Teacher's Name ________________ Class __________ Room _________ Date __________

Lesson Plan

1-day lesson (See *Pacing and Assignment Guide*, TE page 2A)

For use with pages 26–30

GOAL **Use mental math to solve an equation.**

State/Local Objectives ____

____

✓ **Check the items you wish to use for this lesson.**

STARTING OPTIONS

____ Homework Check (1.4): TE page 21; Answer Transparencies

____ Homework Quiz (1.4): TE page 22; Transparencies

____ Warm-Up: Transparencies

TEACHING OPTIONS

____ Notetaking Guide

____ Activity: SE page 26

____ Examples: 1–3, SE pages 26–28

____ Extra Examples: TE pages 27–28

____ Your Turn Now Exercises: 1–8, SE pages 27–28

____ Concept Check: TE page 28

____ Getting Ready to Practice Exercises: 1–11, SE page 28

APPLY/HOMEWORK

Homework Assignment

____ Basic: Day 1: SRH p. 696 Exs. 3, 5; pp. 29–30 Exs. 12–23, 28–31, 33, 41–52

____ Average: Day 1: pp. 29–30 Exs. 12, 14, 17, 18, 23–37, 41 52

____ Advanced: Day 1: pp. 29–30 Exs. 14, 23–48*, 51, 52

Reteaching the Lesson

____ Practice: CRB pages 43–45 (Level A, Level B, Level C); Practice Workbook

____ Study Guide: CRB pages 46–47; Spanish Study Guide

Extending the Lesson

____ Challenge: SE page 30; CRB page 48

ASSESSMENT OPTIONS

____ Daily Quiz (1.5): TE page 30 or Transparencies

____ Test-Taking Practice: SE page 30

Notes

____

____

LESSON 1.5

Teacher's Name ____________ Class ________ Room ________ Date ________

Lesson Plan for Block Scheduling

Half-block lesson (See *Pacing and Assignment Guide*, TE page 2A)

For use with pages 26-30

GOAL **Use mental math to solve an equation.**

State/Local Objectives ____

____

✓ **Check the items you wish to use for this lesson.**

Chapter Pacing Guide	
Day	**Lesson**
1	1.1; 1.2 (begin)
2	1.2 (end); 1.3
3	1.4
4	**1.5;** 1.6
5	1.7
6	Ch. 1 Review and Projects

STARTING OPTIONS

____ Homework Check (1.4): TE page 21; Answer Transparencies

____ Homework Quiz (1.4): TE page 22; Transparencies

____ Warm-Up: Transparencies

TEACHING OPTIONS

____ Notetaking Guide

____ Activity: SE page 26

____ Examples: 1–3, SE pages 26–28

____ Extra Examples: TE pages 27–28

____ Your Turn Now Exercises: 1–8, SE pages 27–28

____ Concept Check: TE page 28

____ Getting Ready to Practice Exercises: 1–11, SE page 28

APPLY/HOMEWORK

Homework Assignment

____ Block Schedule: pp. 29–30 Exs. 12, 14, 17, 18, 23–37, 41–52 (with 1.6)

Reteaching the Lesson

____ Practice: CRB pages 43–45 (Level A, Level B, Level C); Practice Workbook

____ Study Guide: CRB pages 46–47; Spanish Study Guide

Extending the Lesson

____ Challenge: SE page 30; CRB page 48

ASSESSMENT OPTIONS

____ Daily Quiz (1.5): TE page 30 or Transparencies

____ Test-Taking Practice: SE page 30

Notes

____

____

____

____

LESSON 1.5

Name ______________________ Date ________

Practice A

For use with pages 26–30

Complete the statement.

1. $\underline{\quad ? \quad} + 2 = 11$
2. $7 - \underline{\quad ? \quad} = 4$
3. $4 \times \underline{\quad ? \quad} = 28$
4. Distance traveled is equal to the speed times the ___?___.

Match the equation with its solution.

5. $x + 18 = 24$ — A. $x = 5$
6. $12x = 60$ — B. $x = 6$
7. $x - 24 = 18$ — C. $x = 42$
8. $x \div 12 = 5$ — D. $x = 60$

Tell whether the given value of the variable is a solution of the equation.

9. $8a = 48$; $a = 6$
10. $17 + x = 35$; $x = 12$
11. $m - 11 = 17$; $m = 28$
12. $\frac{b}{4} = 36$; $b = 9$
13. $27 - t = 15$; $t = 12$
14. $c + 9 = 38$; $c = 28$

Solve the equation using mental math.

15. $3b = 18$
16. $x - 5 = 7$
17. $32 \div a = 8$
18. $6 + p = 11$
19. $13 - m = 5$
20. $12z = 48$

21. In the 1964 Summer Olympics, Bob Hayes of the United States ran and won the 100-meter race in 10 seconds. What was Bob Hayes' running speed?
22. You bike at a speed of 10 miles per hour. At this speed, how long would it take you to bike 30 miles?
23. Your part-time summer job pays \$6 an hour. Your goal is to make \$120 a week. Solve the equation $6h = 120$, where h is the number of hours worked, to find how many hours you need to work to meet your goal.
24. Two scales used for measuring temperature are the Kelvin scale and the Celsius scale. A temperature in degrees Kelvin is about 273 degrees greater than a temperature in degrees Celsius. Solve the equation $373 = C + 273$ to find the temperature in degrees Celsius that is the same as 373 degrees Kelvin.

LESSON 1.5

Name ______________________________ Date __________

Practice B

For use with pages 26–30

Tell whether the given value of the variable is a solution of the equation.

1. $7x = 42; x = 6$

2. $7 + a = 19; a = 12$

3. $z \div 4 = 6; z = 2$

4. $b - 16 = 28; b = 34$

5. $n + 8 = 21; n = 13$

6. $33 - y = 18; y = 25$

Match the equation with the corresponding question. Then solve.

7. $6x = 48$ **A.** 6 plus what number equals 48?

8. $x \div 6 = 8$ **B.** What number divided by 6 equals 8?

9. $6 + x = 48$ **C.** 6 times what number equals 48?

10. $\frac{48}{x} = 6$ **D.** 48 divided by what number equals 6?

Solve the equation using mental math.

11. $b + 15 = 23$

12. $25 - x = 16$

13. $40 \div a = 8$

14. $7m = 56$

15. $y \div 4 = 10$

16. $p - 12 = 7$

Use the formula for distance to find the unknown value.

17. d = 100 miles, r = __?__, t = 5 hours

18. d = 360 kilometers, r = 40 kilometers per hour, t = __?__

19. d = 420 miles, r = __?__, t = 7 hours

20. The slowest flying bird in the world flies at a speed of 5 miles per hour. How long does it take for the bird to fly 15 miles?

21. Telescopes are described by their f/number. The f/number is given by the equation $f = F \div D$, where f is the f/number, F is the focal length of the telescope, and D is the width of the telescope lens. Use the f/number equation to find the focal length (in inches) of a telescope with a 10-inch wide lens and an f/number of 9.

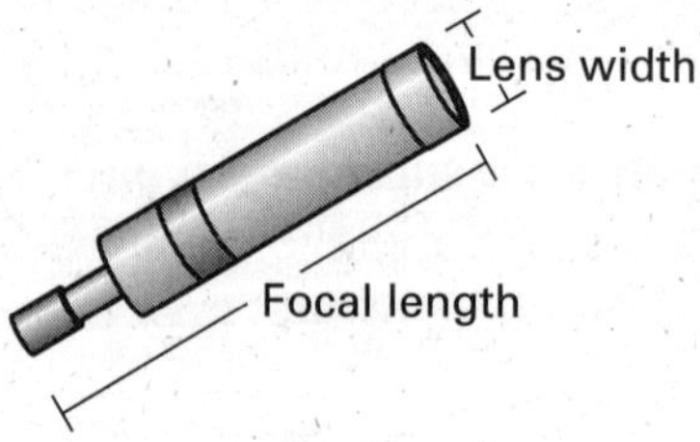

22. Two scales used for measuring temperature are the Kelvin scale and the Celsius scale. A temperature in degrees Kelvin is about 273 degrees greater than a temperature in degrees Celsius. Solve the equation $300 = C + 273$ to find the temperature in degrees Celsius that is the same as 300 degrees Kelvin.

LESSON 1.5

Name ______________________________ Date __________

Practice C

For use with pages 26–30

Lesson 1.5

Tell whether the given value of the variable is a solution of the equation.

1. $4a = 64; a = 16$

2. $15 + y = 29; y = 13$

3. $39 \div b = 3; b = 13$

4. $x - 11 = 23; x = 32$

5. $m + 31 = 57; m = 26$

6. $z \div 21 = 3; z = 7$

Solve the equation using mental math.

7. $r + 16 = 25$

8. $\frac{110}{x} = 10$

9. $31 - a = 12$

10. $12n = 72$

11. $9 + b = 24$

12. $z - 14 = 18$

Use the formula for distance to find the unknown value.

13. $d = 200$ miles, $r = 50$ miles per hour, $t =$ __?__

14. $d = 164$ miles, $r =$ __?__, $t = 4$ hours

15. $d = 500$ kilometers, $r = 25$ kilometers per hour, $t =$ __?__

16. Sound travels at different speeds through different media (air, water, metal, etc.). For example, sound travels faster through iron than through air. If sound travels through iron at a speed of 3 miles per second, how long does it take sound to travel through 12 miles of iron?

17. Telescopes are described by their f/number. The f/number is given by the equation $f = F \div D$, where f is the f/number, F is the focal length of the telescope, and D is the width of the telescope lens. Use the f/number equation to find the focal length (in inches) of a telescope with a 10-inch wide lens and an f/number of 15.

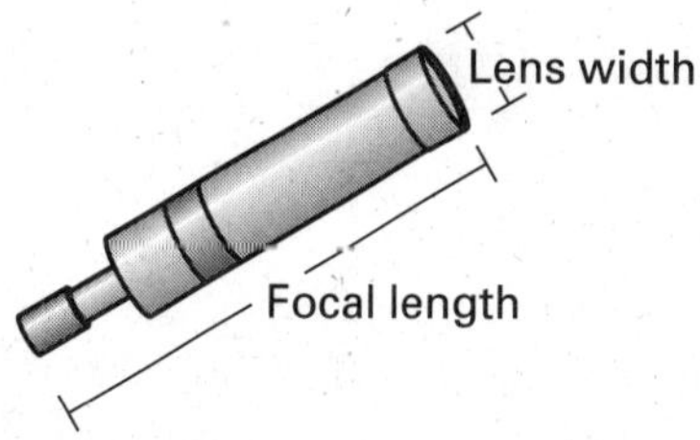

18. Two scales used for measuring temperature are the Kelvin scale and the Celsius scale. A temperature in degrees Kelvin is about 273 degrees greater than a temperature in degrees Celsius. Solve the equation $450 = C + 273$ to find the temperature in degrees Celsius that is the same as 450 degrees Kelvin.

19. Write an equation for the following question. Then solve the equation.

Ninety-six divided by what number equals 32?

Solve the equation using mental math.

20. $\frac{15}{x} = 1$

21. $2a + 3 = 13$

22. $3m - 1 = 17$

Lesson 1.5

LESSON 1.5

Name ______________________________ Date ____________

Study Guide

For use with pages 26–30

GOAL **Use mental math to solve an equation.**

VOCABULARY

An **equation** is a mathematical sentence formed by setting two expressions equal.

A **solution** of an equation is a number that you can substitute for the variable to make the equation true.

Finding all solutions of an equation is called **solving the equation.**

The Distance Formula:

Words: Distance traveled is equal to the speed (rate of travel) times the travel time.

Algebra: $d = rt$

EXAMPLE 1 Checking Possible Solutions

Tell whether the value of the variable is a solution of $18 - h = 7$.

a. $h = 11$ **b.** $h = 9$

Solution

a.	$18 - h = 7$	Write equation.
	$18 - \mathbf{11} \stackrel{?}{=} 7$	Substitute 11 for h.
	$7 = 7$ ✓	The equation is true, so 11 is a solution.
b.	$18 - h = 7$	Write equation.
	$18 - \mathbf{9} \stackrel{?}{=} 7$	Substitute 9 for h.
	$9 \neq 7$	The equation is not true, so 9 is not a solution.

Exercises for Example 1

Tell whether the value of the variable is a solution of the equation.

1. $b \div 4 = 7; b = 28$ **2.** $8w = 56; w = 6$ **3.** $6 + c = 11; c = 5$

Name ______________________________ Date __________

Study Guide

For use with pages 26–30

EXAMPLE 2 Using Mental Math to Solve Equations

	Equation	Question	Solution	Check
a.	$m + 14 = 20$	What number plus 14 equals 20?	6	$6 + 14 = 20$
b.	$9 - k = 2$	9 minus what number equals 2?	7	$9 - 7 = 2$
c.	$6x = 24$	6 times what number equals 24?	4	$6(4) = 24$
d.	$63 \div r = 7$	63 divided by what number equals 7?	9	$63 \div 9 = 7$

Exercises for Example 2

Solve the equation using mental math.

4. $t - 8 = 12$ **5.** $5q = 30$ **6.** $17 + p = 25$ **7.** $d \div 3 = 9$

EXAMPLE 3 Using Mental Math to Solve an Equation

A green sea turtle can swim at a speed of about 20 miles per hour. About how long would it take for a green sea turtle to swim 80 miles?

Solution

$d = rt$	Write the formula for distance.
$80 = 20t$	Substitute the values you know.
$80 = 20 \cdot 4$	Use mental math to solve the equation.

Answer: It would take a green sea turtle about 4 hours to swim 80 miles.

Exercise for Example 3

8. A runner runs a race at a steady pace. If it takes the runner 42 minutes to run 6 miles, how long did it take the runner to run each mile?

LESSON 1.5

Name ______________________ Date __________

Challenge Practice

For use with pages 26-30

Solve the equation using mental math.

1. $\frac{k}{2} - 6 = 5$

2. $\frac{m - 6}{2} = 5$

3. $83 + 11x = 83$

4. $24 - 5b = 9$

5. $2(z + 1) = 100$

6. $5(s - 4) = 60$

7. $5(r + 18) = 125$

8. $6p - 19 = 35$

9. You and your younger brother walk to school every morning. You walk at a rate of 4 feet per second while your brother walks at a rate of 3 feet per second. If you both leave from home at the same time and the school is 2400 feet from your house, how long after you does your brother arrive at school?

LESSON 1.6

Teacher's Name ________________ Class __________ Room _________ Date __________

Lesson Plan

1-day lesson (See *Pacing and Assignment Guide*, TE page 2A)

For use with pages 31–35

GOAL **Use formulas to find perimeter and area.**

State/Local Objectives ____

____

✓ Check the items you wish to use for this lesson.

STARTING OPTIONS

____ Homework Check (1.5): TE page 29; Answer Transparencies

____ Homework Quiz (1.5): TE page 30; Transparencies

____ Warm-Up: Transparencies

TEACHING OPTIONS

____ Notetaking Guide

____ Hands-on Activity: SE page 31

____ Examples: 1–3, SE pages 32–33

____ Extra Examples: TE page 33

____ Your Turn Now Exercises: 1–5, SE page 33

____ Concept Check: TE page 33

____ Getting Ready to Practice Exercises: 1–7, SE page 34

APPLY/HOMEWORK

Homework Assignment

____ Basic: Day 1: SRH p. 696 Exs. 1, 2, 4; pp. 34–35 Exs. 8–18, 20, 27–35

____ Average: Day 1: pp. 34–35 Exs. 9, 12–23, 27–35

____ Advanced: Day 1: pp. 34–35 Exs. 9, 14–30*, 34, 35

Reteaching the Lesson

____ Practice: CRB pages 51–53 (Level A, Level B, Level C); Practice Workbook

____ Study Guide: CRB pages 54–55; Spanish Study Guide

Extending the Lesson

____ Real-World Problem Solving: CRB page 56

____ Challenge: SE page 35; CRB page 57

ASSESSMENT OPTIONS

____ Daily Quiz (1.6): TE page 35 or Transparencies

____ Test-Taking Practice: SE page 35

Notes

____

____

LESSON 1.6

Teacher's Name ________________ Class __________ Room __________ Date __________

Lesson Plan for Block Scheduling

Half-block lesson (See *Pacing and Assignment Guide*, TE page 2A)

For use with pages 31–35

GOAL **Use formulas to find perimeter and area.**

State/Local Objectives ____

____

Lesson 1.6

✓ Check the items you wish to use for this lesson.

Chapter Pacing Guide	
Day	**Lesson**
1	1.1; 1.2 (begin)
2	1.2 (end); 1.3
3	1.4
4	1.5; **1.6**
5	1.7
6	Ch. 1 Review and Projects

STARTING OPTIONS

____ Homework Check (1.5): TE page 29; Answer Transparencies

____ Homework Quiz (1.5): TE page 30; Transparencies

____ Warm-Up: Transparencies

TEACHING OPTIONS

____ Notetaking Guide

____ Hands-on Activity: SE page 31

____ Examples: 1–3, SE pages 32–33

____ Extra Examples: TE page 33

____ Your Turn Now Exercises: 1–5, SE page 33

____ Concept Check: TE page 33

____ Getting Ready to Practice Exercises: 1–7, SE page 34

APPLY/HOMEWORK

Homework Assignment

____ Block Schedule: pp. 34–35 Exs. 9, 12–23, 27–35 (with 1.5)

Reteaching the Lesson

____ Practice: CRB pages 51–53 (Level A, Level B, Level C); Practice Workbook

____ Study Guide: CRB pages 54–55; Spanish Study Guide

Extending the Lesson

____ Real-World Problem Solving: CRB page 56

____ Challenge: SE page 35; CRB page 57

ASSESSMENT OPTIONS

____ Daily Quiz (1.6): TE page 35 or Transparencies

____ Test-Taking Practice: SE page 35

Notes

____

____

____

LESSON 1.6

Name ______________________ Date __________

Practice A

For use with pages 31–35

Evaluate the power.

1. 3^2 **2.** 10^2 **3.** 8^2

Complete the statement.

4. The __?__ of the length and width of a rectangle gives the area of the rectangle.

5. Which expression gives the area of the rectangle?

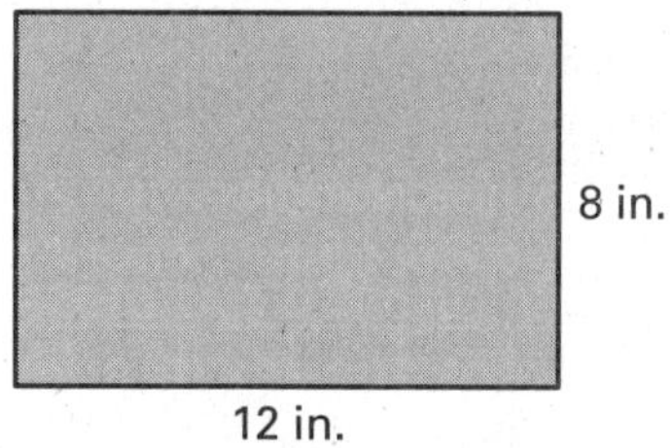

A. $2(12) + 2(8)$

B. $4(12)$

C. $12(8)$

D. 12^2

Find the perimeter and the area of the rectangle or square with the given dimensions.

6.

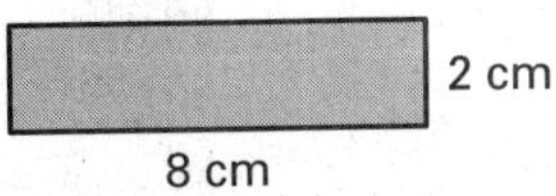

7.

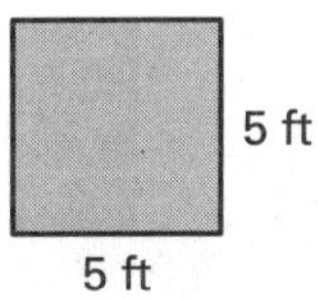

8. $\ell = 7$ yards, $w = 6$ yards **9.** $s = 9$ meters

10. A computer screen is 13 inches long and 10 inches wide. Find the perimeter and the area of the computer screen.

11. The standard size for a photograph is 5 inches wide and 7 inches long. What are the perimeter and the area of a standard-sized photograph?

12. You plant a rectangular flower bed that has a length of 12 feet and a width of 5 feet. What are the perimeter and the area of the flower bed?

13. The tablet that the Statue of Liberty is holding is about 24 feet long and 14 feet wide. What are the perimeter and the area of the Statue of Liberty's tablet?

Name ______________________________ Date ________

Practice B

For use with pages 31–35

Match the figure with its area or perimeter.

1.

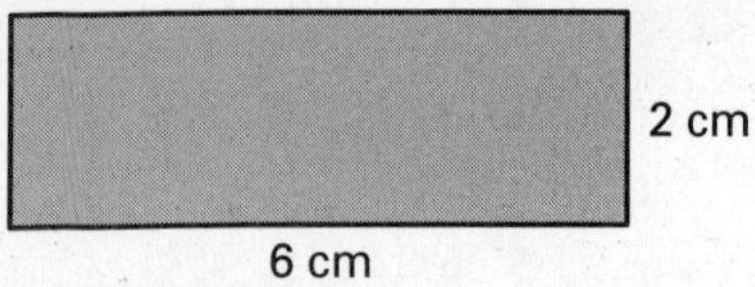

2.

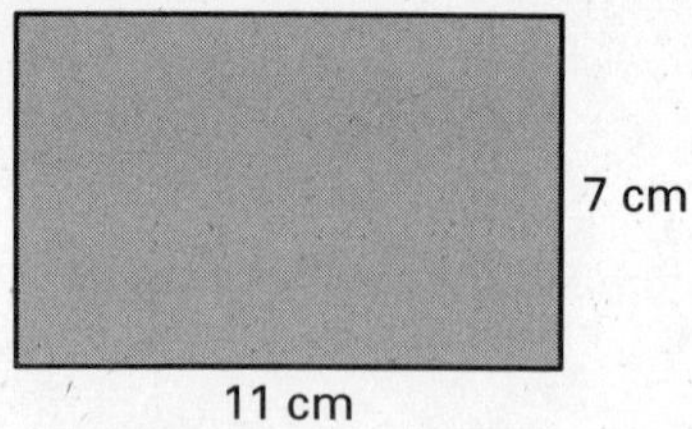

3.

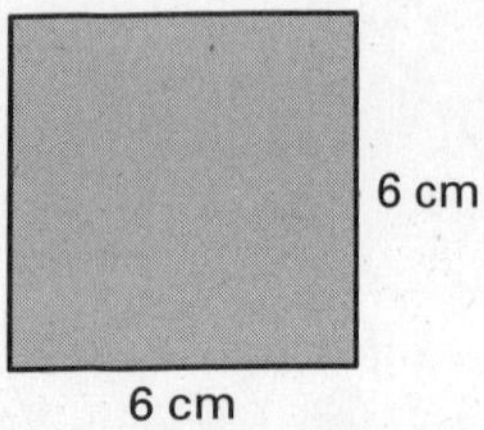

4.

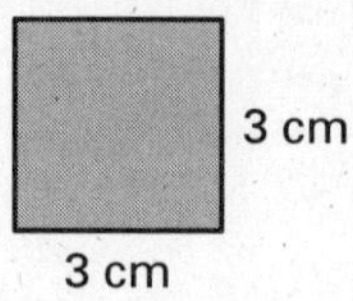

A. 12 centimeters

B. 12 square centimeters

C. 36 centimeters

D. 36 square centimeters

Find the perimeter and the area of the rectangle or square with the given dimensions.

5. $\ell = 11$ inches, $w = 5$ inches

6. $s = 4$ meters

7. $s = 10$ millimeters

8. $\ell = 13$ feet, $w = 2$ feet

9. $\ell = 8$ centimeters, $w = 7$ centimeters

10. $s = 7$ inches

11. $s = 3$ meters

12. $\ell = 12$ yards, $w = 6$ yards

13. Which two rectangles with the given length and width have the same area?

A.

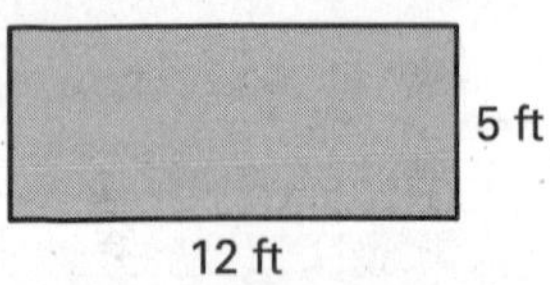

B.

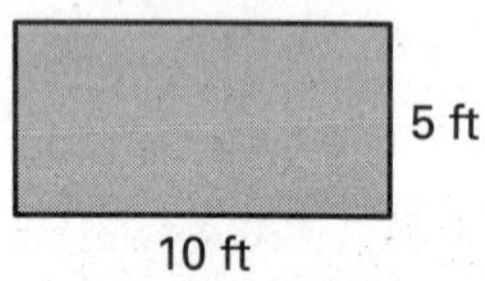

C.

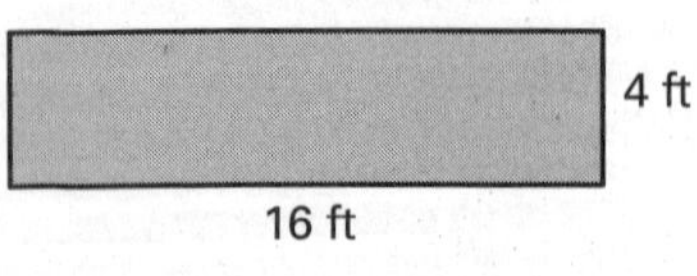

D.

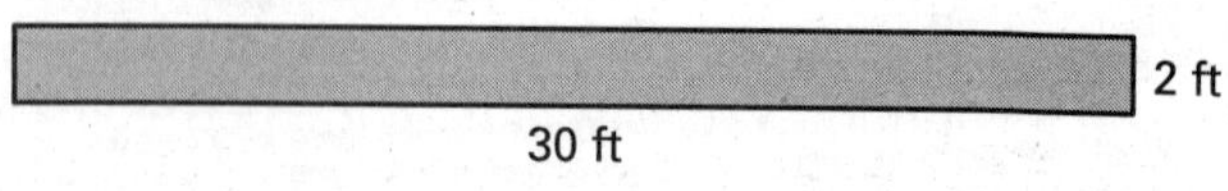

14. A hand held video game has a screen that is about 61 millimeters long and 41 millimeters wide. What is the perimeter and the area of the screen?

15. Measure the sides of the rectangle to the nearest centimeter. Use your measurements to find the perimeter and the area of the rectangle.

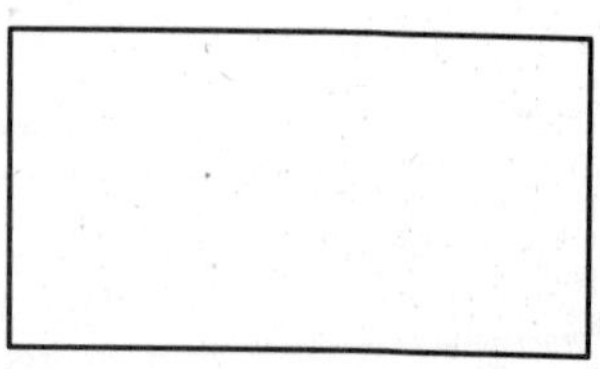

Name ______________________________ Date __________

Practice C

For use with pages 31–35

Find the perimeter and the area of the rectangle or square with the given dimensions.

1. $\ell = 12$ inches, $w = 9$ inches

2. $s = 8$ meters

3. $s = 6$ millimeters

4. $\ell = 12$ feet, $w = 11$ feet

5. $\ell = 30$ centimeters, $w = 20$ centimeters

6. $s = 15$ inches

7. $s = 14$ meters

8. $\ell = 15$ yards, $w = 8$ yards

9. Which two rectangles with the given length and width have the same perimeter?

10. A college football field is a rectangle with a length of 360 feet and a width of 160 feet. What are the perimeter and the area of the football field?

11. What is the difference in area between a computer screen that is 12 inches long and 9 inches wide and a computer screen that is 11 inches long and 8 inches wide?

12. Name the length and width of two different rectangles that could have an area of 16 square feet.

13. A rectangle is 2 inches longer than it is wide. If the width is 9 inches, what are the perimeter and the area of the rectangle?

14. The figure below can be broken into rectangles. Find the perimeter and the area of the figure. Explain your method.

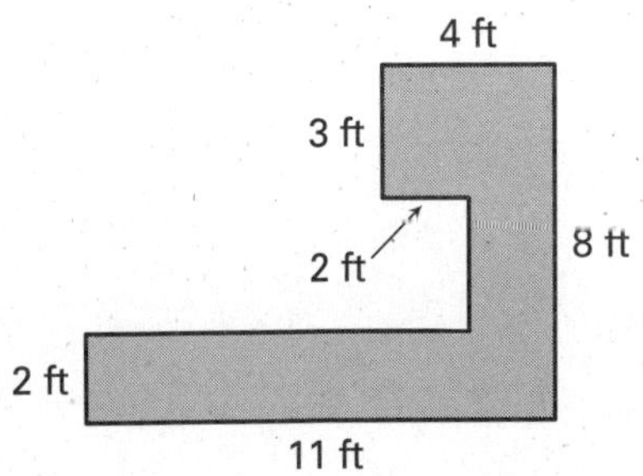

LESSON 1.6

Name ______________________ Date __________

Study Guide

For use with pages 31–35

GOAL **Use formulas to find perimeter and area.**

VOCABULARY

The **perimeter** of a rectangle is the sum of the lengths of the sides.

Rectangle: $P = 2\ell + 2w$ Square: $P = 4s$

The **area** of a rectangle is the number of square units needed to cover the rectangle.

Rectangle: $A = \ell w$ Square: $A = s^2$

EXAMPLE 1 Finding Perimeter

Find the perimeter of the rectangle or square with the given dimensions.

a. $s = 6$ inches **b.** $\ell = 7$ meters, $w = 3$ meters

Solution

a. $P = 4s$ Write formula.

$= 4(\mathbf{6})$ Substitute 6 for s.

$= 24$ Multiply.

Answer: The perimeter is 24 inches.

b. $P = 2\ell + 2w$ Write formula.

$= 2(\mathbf{7}) + 2(\mathbf{3})$ Substitute 7 for ℓ and 3 for w.

$= 14 + 6$ Multiply.

$= 20$ Add.

Answer: The perimeter is 20 meters.

Exercises for Example 1

Find the perimeter of the rectangle or square.

1. 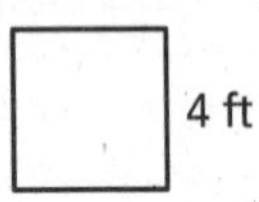

2. 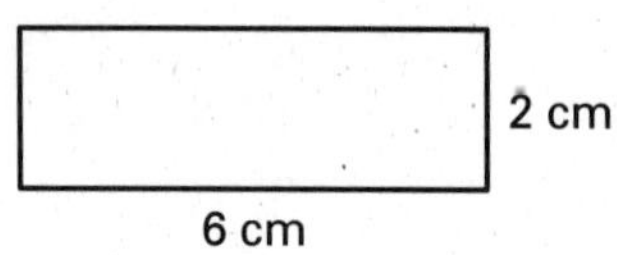

Name ______________________________ Date __________

Study Guide

For use with pages 31–35

EXAMPLE 2 Finding Area

Find the area of the rectangle or square.

a.

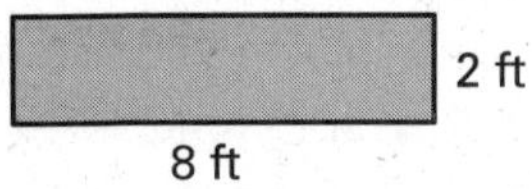

b.

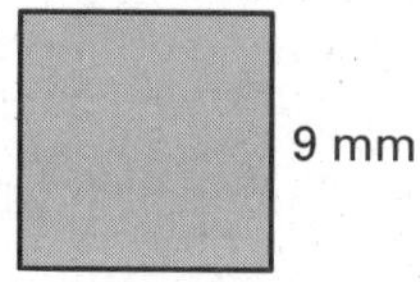

Solution

a. $A = \ell w$
$= 8(2)$
$= 16$

Answer: The area is 16 ft^2.

b. $A = s^2$
$= 9^2$
$= 81$

Answer: The area is 81 mm^2.

Exercises for Example 2

Find the area of the rectangle or square with the given dimensions.

3. $\ell = 8$ meters, $w = 4$ meters

4. $s = 7$ yards

5. Find the perimeter and the area of a rectangle that has a length of 9 centimeters and a width of 6 centimeters.

EXAMPLE 3 Using Perimeter and Area

International soccer is played on a rectangular field that has a length of 100 meters and a width of 64 meters. What is the perimeter and the area of the soccer field?

Solution

$P = 2\ell + 2w$
$= 2(100) + 2(64)$
$= 200 + 128$
$= 328$

$A = \ell w$
$= 100(64)$
$= 6400$

Answer: The perimeter of the soccer field is 328 meters, and the area is 6400 square meters.

Exercise for Example 3

6. A picture frame holds a picture that has a length of 6 inches and a width of 4 inches. What is the perimeter and the area of the picture?

Name ______________________ Date ________

Real-World Problem Solving

For use with pages 31–35

Home Improvement

Any house, after time, will start needing repairs. Some can be minor improvements like painting a wall or replacing a faucet; others can be costly and time consuming like remodeling a kitchen or replacing a roof. Ignoring areas of repair can cause a house to lose its value or even worse, make the house hazardous to live in. Home improvements can also mean making changes to the area around a house. Replacing a porch or planting shrubs by the house can be considered home improvement.

In Exercises 1–4, use the following information.

John and Mary are buying a house. Before moving into the house they are planning on making some improvements. The improvements are carpeting 5 rooms, fencing the backyard, and paving the driveway. To find the cost for the improvements John and Mary need to calculate the area of each room for the carpeting, the perimeter of the backyard for the fencing, and the area of the driveway for paving. The measurements are shown below.

Carpeting: Room 1: $\ell = 12$ ft, $w = 9$ ft
Room 2: $\ell = 14$ ft, $w = 10$ ft
Room 3: $\ell = 12$ ft, $w = 11$ ft
Room 4: $\ell = 12$ ft, $w = 10$ ft
Room 5: $\ell = 14$ ft, $w = 9$ ft

Driveway: $\ell = 60$ ft, $w = 15$ ft

Backyard:

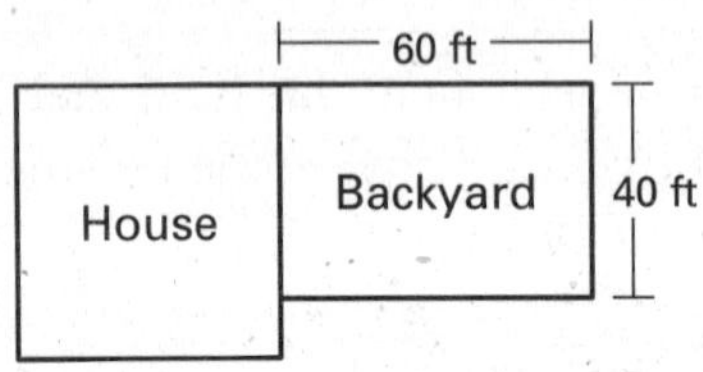

1. What is the area of each room?
2. What is the area of the driveway?
3. What is the perimeter of the backyard?
4. If John and Mary do not want fencing against the house, how many feet of fencing do they need?

LESSON 1.6

Name ____________________ Date __________

Challenge Practice

For use with pages 31–35

1. Find the perimeter and area of the figure.

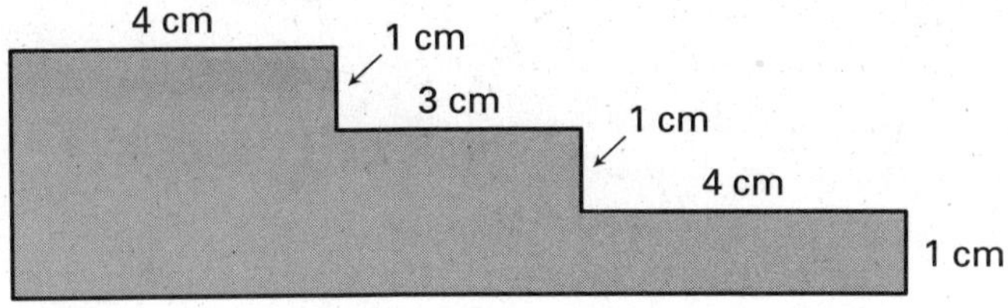

2. A square has a perimeter of 144 meters. What is the area of the square?

3. A square has an area of 144 square meters. What is the perimeter of the square?

4. A rectangle has a perimeter of 64 inches. How long are the sides of a square having the same perimeter as the rectangle? What is the area of the square?

In Exercises 5–7, use the following information. A rectangular bulletin board requires 120 square feet of paper to cover it.

5. Fill in the table with a list of 5 different sets of possible dimensions for this bulletin board.

Length	Width

6. If the bulletin board is 24 feet long, how wide would it be?

7. Determine the dimensions of the bulletin board if the perimeter is 52 feet.

Lesson 1.6

LESSON 1.7

Teacher's Name ________________ Class __________ Room _________ Date __________

Lesson Plan

2-day lesson (See *Pacing and Assignment Guide*, TE page 2A)
For use with pages 36–41

GOAL **Use a 4-step plan to solve many kinds of problems.**

State/Local Objectives ____

____

✓ Check the items you wish to use for this lesson.

STARTING OPTIONS

____ Homework Check (1.6): TE page 34; Answer Transparencies
____ Homework Quiz (1.6): TE page 35; Transparencies
____ Warm-Up: Transparencies

TEACHING OPTIONS

____ Notetaking Guide
____ Problem Solving Strategies: SE pages 36–37
____ Examples: Day 1: 1–2, SE pages 38–39; Day 2: none
____ Extra Examples: TE page 39
____ Your Turn Now Exercises: Day 1: 1–3, SE pages 38–39; Day 2: none
____ Concept Check: TE page 39
____ Getting Ready to Practice Exercises: Day 1: 1, SE page 40; Day 2: 2, SE page 40

APPLY/HOMEWORK

Homework Assignment

____ Basic: Day 1: SRH p. 697 Exs. 1–6; pp. 40–41 Exs. 3–9 odd, 15–21 odd
Day 2: pp. 40–41 Exs. 4–10 even, 14–20 even

____ Average: Day 1: pp. 40–41 Exs. 3–11 odd, 15–21 odd
Day 2: pp. 40–41 Exs. 4–20 even

____ Advanced: Day 1: pp. 40–41 Exs. 3–21* odd
Day 2: pp. 40–41 Exs. 4–20 even

Reteaching the Lesson

____ Practice: CRB pages 60–62 (Level A, Level B, Level C); Practice Workbook
____ Study Guide: CRB pages 63–64; Spanish Study Guide

Extending the Lesson

____ Challenge: SE page 41; CRB page 65

ASSESSMENT OPTIONS

____ Daily Quiz (1.7): TE page 41 or Transparencies
____ Test-Taking Practice: SE page 41
____ Quiz (1.5–1.7): SE page 43; Assessment Book page 8

Notes ____

LESSON 1.7

Teacher's Name ______________ Class __________ Room __________ Date __________

Lesson Plan for Block Scheduling

1-block lesson (See *Pacing and Assignment Guide,* TE page 2A)

For use with pages 36–41

GOAL **Use a 4-step plan to solve many kinds of problems.**

State/Local Objectives ____

____

✓ Check the items you wish to use for this lesson.

Chapter Pacing Guide	
Day	**Lesson**
1	1.1; 1.2 (begin)
2	1.2 (end); 1.3
3	1.4
4	1.5; 1.6
5	**1.7**
6	Ch. 1 Review and Projects

STARTING OPTIONS

____ Homework Check (1.6): TE page 34; Answer Transparencies

____ Homework Quiz (1.6): TE page 35; Transparencies

____ Warm-Up: Transparencies

TEACHING OPTIONS

____ Notetaking Guide

____ Problem Solving Strategies: SE pages 36–37

____ Examples: 1–2, SE pages 38–39

____ Extra Examples: TE page 39

____ Your Turn Now Exercises: 1–3, SE pages 38–39

____ Concept Check: TE page 39

____ Getting Ready to Practice Exercises: 1–2, SE page 40

APPLY/HOMEWORK

Homework Assignment

____ Block Schedule: pp. 40–41 Exs. 3–12, 14–21

Reteaching the Lesson

____ Practice: CRB pages 60–62 (Level A, Level B, Level C); Practice Workbook

____ Study Guide: CRB pages 63–64; Spanish Study Guide

Extending the Lesson

____ Challenge: SE page 41; CRB page 65

ASSESSMENT OPTIONS

____ Daily Quiz (1.7): TE page 41 or Transparencies

____ Test-Taking Practice: SE page 41

____ Quiz (1.5–1.7): SE page 43; Assessment Book page 8

Notes

____

____

____

LESSON 1.7

Name ______________________________ Date __________

Practice A

For use with pages 36–41

You have $100 to spend on a new outfit. You spend $25 on shoes and you still need pants and a shirt. You find 2 pants and 2 shirts that you like. The prices for the items are shown in the table.

Item	Price
Tan pants	$35
Navy pants	$40
Striped shirt	$36
Solid shirt	$30

1. How much did you spend on shoes?

2. How much money does this leave you to spend on pants and a shirt?

3. Which combinations of pants and a shirt could you buy with the money you have left?

In Exercises 4–7, identify the operation (addition, subtraction, multiplication, or division) that needs to be used in order to solve the problem.

Lesson 1.7

4. One serving of rice weighs 48 grams. If a bag of rice weighs 2260 grams, how many full servings of rice are in the bag?

5. Your friend lives 3 miles farther away from school than you do. If you live 12 miles from school, how far away from school does your friend live?

6. If one T-shirt costs $10, how much do 15 T-shirts cost?

7. Your brother is 5 inches shorter than you. If you are 67 inches tall, how tall is your brother?

8. The sum of the digits of a two digit number is 8. The tens' digit is 2 less than the ones' digit. What is the number?

9. You have two bedrooms in your home that are located side by side, as shown in the figure. You are having the same carpeting installed in both bedrooms. How much carpeting do you need for both bedrooms?

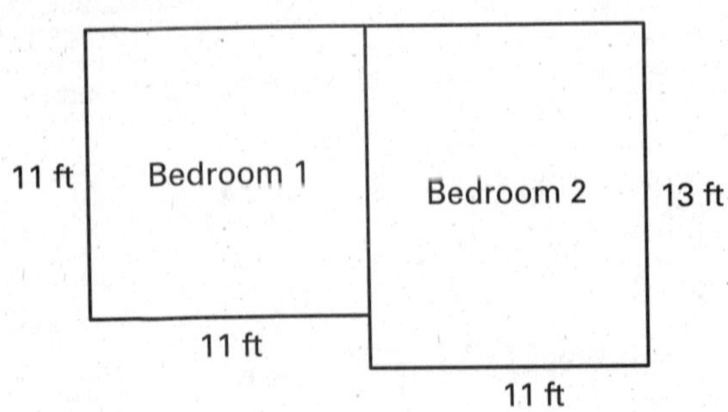

LESSON 1.7

Name ______________________ Date __________

Practice B

For use with pages 36–41

1. Describe and correct the error made in solving the following problem.

You are ordering pizza for a party. One pizza serves 8 people. You expect that there will be 35 people at the party. How many pizzas should you order?

$$\begin{array}{r} 4\text{ R3} \\ 8\overline{)35} \\ \underline{32} \\ 3 \end{array}$$

You need to order 4 pizzas.

2. Farmers typically do not plant the same crop in the same plot of soil season after season. For each upcoming season, the farmer will switch positions of crops from the previous season to grow healthier plants. This is called crop rotation. The tables below show the rotation of crops for three seasons. If you continue the pattern, how many seasons will it take for the crops to be in the positions they were in the first season?

Season 1		
Beans	Potatoes	Lettuce
Cabbage	Spinach	Carrots

Season 2		
Cabbage	Beans	Potatoes
Spinach	Carrots	Lettuce

Season 3		
Spinach	Cabbage	Beans
Carrots	Lettuce	Potatoes

3. You have \$100 to spend on a new outfit. You spend \$25 on shoes and you still need pants and a shirt. You find 3 pants and 2 shirts that you like. The prices for the items are shown in the table. Which combinations of pants and a shirt could you buy with the money you have left?

Item	Price
Tan pants	\$48
Navy pants	\$34
Black pants	\$42
Striped shirt	\$38
Solid shirt	\$32

4. A computer screen is divided into little squares called pixels. The number of pixels on the screen is called the monitor resolution. A common monitor resolution is 640×480, which means that there are 640 pixels along the length of the screen and 480 pixels along the width of the screen. Another common monitor resolution is 800×600. How many pixels is the computer screen divided into with each resolution? Which resolution uses more pixels and how many more pixels does it use?

5. A pizza shop offers five different toppings (pepperoni (R), onions (O), peppers (P), sausage (S), and mushrooms (M)) for its pizzas. How many different pizzas can be created using three different toppings? What are the combinations of the toppings?

LESSON 1.7

Name ______________________ Date ________

Practice C

For use with pages 36–41

Lesson 1.7

1. What should you do if you solve a problem and look back to find that your answer is not reasonable?

2. Copy and complete the table for the powers of 7. Without actually calculating 7^7, how can you tell what the last digit is?

Power	7^1	7^2	7^3	7^4	7^5	7^6
Value	7	49	343	?	?	?

3. You have \$105 to spend on a new outfit. You spend \$25 on shoes and you still need pants and a shirt. You find 3 pants and 3 shirts that you like. The prices for the items are shown in the table. Which combinations of pants and a shirt could you buy with the money you have left?

Item	**Price**
Tan pants	\$46
Navy pants	\$33
Black pants	\$35
White shirt	\$42
Solid shirt	\$32
Striped shirt	\$38

4. A computer screen is divided into little squares called pixels. The number of pixels on the screen is called the monitor resolution. A common monitor resolution is 640 × 480, which means that there are 640 pixels along the length of the screen and 480 pixels along the width of the screen. Another common monitor resolution is 1024 × 768. How many pixels is the computer screen divided into with each resolution? Which resolution uses more pixels and how many more pixels does it use?

5. A farmer is enclosing a pasture into two areas as shown in the figure below. Find the total area of the pasture and the amount of fencing needed to enclose the pasture into the two areas.

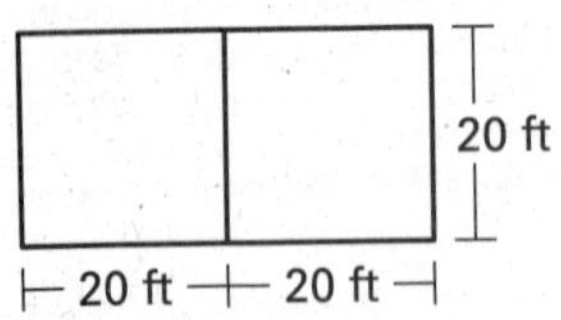

6. The sum of the heights, in inches, of Judy, Keith, and Sara is 192 inches. If Judy is one inch shorter than Keith and Sara is one inch taller than Keith, how tall are Judy, Keith, and Sara?

LESSON 1.7

Name ______________________________ Date __________

Study Guide

For use with pages 36–41

GOAL **Use a 4-step plan to solve many kinds of problems.**

Problem Solving Plan

1. **Read and Understand** Read the problem carefully. Identify the question and any important information.
2. **Make a Plan** Decide on a problem solving strategy.
3. **Solve the Problem** Use the problem solving strategy to answer the question.
4. **Look Back** Check that your answer is reasonable.

EXAMPLE 1 Understanding and Planning

Your parents say that you can spend at most \$200 on activities for the summer. You can choose 3 activities from the table. You know you want to go to the amusement park. What activities can you plan for the summer?

Summer Activity	Cost
Amusement park	\$42
Pass to pool	\$40
Zoo membership	\$45
Basketball camp	\$145
Camping trip	\$75

Read and Understand

What do you know?

You can spend at most \$200.

The table gives the cost of the activities.

You want to choose 3 activities, and one is the amusement park.

What do you want to find out?

What combinations of activities can you plan for the summer?

Make a plan

How can you relate what you know to what you want to find out?

Find the amount of money available for 2 activities that are not the amusement park.

List all possible combinations of 2 activities and the cost to do them.

Identify the combinations that fit within the available budget.

Exercises for Example 1

1. How much does going to the amusement park cost?
2. How much money does this leave for the other two activities?

LESSON 1.7 Continued

Name ______________________ Date ________

Study Guide

For use with pages 36–41

EXAMPLE 2 Solving and Looking Back

To solve the summer activities problem from Example 1, you need to carry out the plan and then check the answer.

Solve the Problem

Because you know you want to go to the amusement park, there is $\$200 - \$42 = \$158$ for the other two activities. Make a list of all the combinations of two activities and the cost to do them.

Activities	Total Cost
Pool: $40, Zoo: $45	$85
Pool: $40, Basketball: $145	$185
Pool: $40, Camping: $75	$115
Zoo: $45, Basketball: $145	$190
Zoo: $45, Camping: $75	$120
Basketball: $145, Camping: $75	$220

Answer: You can go to the amusement park and either the pool and zoo, the pool and camping, or the zoo and camping.

Look Back

Since basketball camp is the most expensive activity, it makes sense that it does not appear in the answer. So, the answer seems reasonable.

Exercise for Example 2

3. Suppose the camping trip and the basketball camp each costs $115. Now what activities can you plan for the summer?

Lesson 1.7

LESSON 1.7

Name ______________________ Date __________

Challenge Practice

For use with pages 36–41

Use any strategy to solve the problem.

1. Without actually calculating 7^{10}, determine what the last digit of this number is.
2. A farmer owns sheep and chickens. You are informed that there are a total of 19 heads and 60 feet. Determine the number of sheep and the number of chickens the farmer owns.
3. The sum of the digits of a three digit number is 14. The one's digit is equal to the ten's digit. The hundred's digit is five more than the one's digit. What is the number?
4. You have 60 yards of fence available to construct a dog run in the shape of a rectangle. What dimensions should you make the dog run to create the largest possible area for your dog?
5. A do-it-yourself car wash charges $1.75 for the first 5 minutes and $.75 for each additional minute. If you have $5 to spend, how long will you have to wash the car?
6. You leave your house for the park walking at a rate of 264 feet per minute. Your sister leaves for the park 10 minutes later on her bike, and rides at a rate of 440 feet per minute. If the park is one mile from your house, will you or your sister get to the park first? Explain. (Note: 1 mile = 5280 feet)

Name ______________________ Date __________

CHAPTER 1

Chapter Review Games and Activities

For use after Chapter 1

Choose the correct answer to each question, then follow the directions to get through the maze.

1. The __?__ of a rectangle is the sum of all four sides.

A. perimeter: move 4 spaces south

B. area: move 2 spaces south

2. A(n) __?__ is a mathematical sentence formed by setting two expressions equal.

A. variable expression: move 5 spaces east

B. equation: move 2 spaces east

3. What are the next three numbers in the pattern: 1, 8, 15, 22, . . . ?

A. 29, 36, 43: move 4 spaces south

B. 28, 35, 42: move 2 spaces east

4. Evaluate the expression $7x$ when $x = 5$.

A. 42: move 2 spaces south

B. 35: move 3 spaces south

5. Evaluate the expression $17 + t$ when $t = 5$.

A. 22: move 10 spaces east

B. 12: move 7 spaces east

6. Write $4 \cdot 4 \cdot 4 \cdot 4 \cdot 4$ as a power.

A. 4^4: move 5 spaces north

B. 4^5: move 7 spaces north

7. Evaluate 8^5.

A. 32,768: move 4 spaces east

B. 4096: move 2 spaces east

8. Solve the equation $\frac{x}{4} = 9$.

A. 28: move 3 spaces south

B. 36: move 2 spaces north

9. What is the area of the square: $s = 5$ inches?

A. 25 square inches: move 3 spaces east

B. 20 square inches: move 2 spaces east

CHAPTER 1

Name ______________________ Date __________

Real-Life Project: Cost of Installing Floors

For use after Chapter 1

Objective **Discover how perimeter and area are used when installing the floors in a home.**

Materials pencil, paper

Investigation *Getting Going* Your uncle is building a house. He wants to order the materials for the floors. The figure shows the floor plan of the house. As you can see, the rooms are squares and rectangles. Your uncle needs to order a floor covering and a type of baseboard for every room in the house. Carpet, linoleum, and tile are his choices for the floor covering and chestnut or natural wood are his choices for the baseboard. Your uncle has $4000 to spend on the flooring materials for the entire house. Use the figure to answer the following questions.

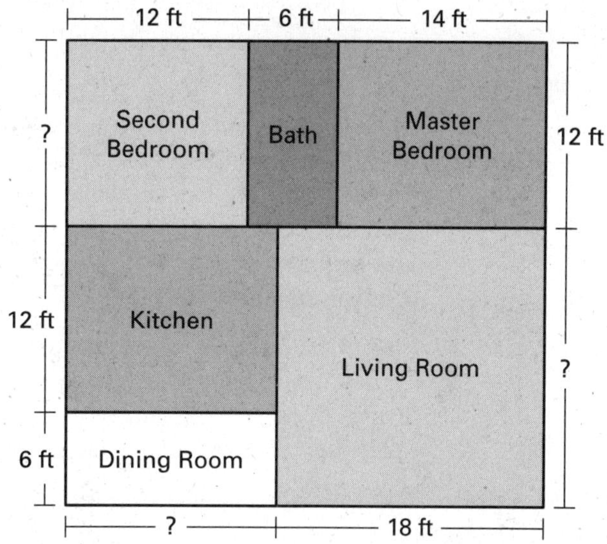

Questions

1. Complete the missing dimensions in the figure. What is the total area of the floor of the house? What is the perimeter?

2. The table below shows the flooring materials and their prices. Find three different ways your uncle can choose to put the flooring materials in the house without going over his budget. Use each flooring material at least once and use each type of baseboard at least once. Every room needs exactly one floor covering to cover the area of each room and exactly one type of baseboard around the perimeter of each room. List the materials that go into each room, the cost for each room, and the cost for the house for each of the three scenarios.

Material		Price
Floor covering	Carpet	$2 per square foot
	Linoleum	$3 per square foot
	Tile	$5 per square foot
Baseboard	Chestnut	$2 per foot
	Natural	$3 per foot

Review and Projects

CHAPTER 1 Continued

Name ______________________ Date __________

Teacher's Notes for Cost of Installing Floors Project

For use after Chapter 1

Project Goals

- Add and multiply whole numbers.
- Solve equations using mental math.
- Use the order of operations with multiplication and addition.
- Use perimeter and area formulas for rectangles and squares.
- Develop problem solving skills.

Managing the Project

Guiding Students' Work Suggest that the students make a table to calculate the costs of the flooring for each room. The table could have the following headings: room, area, perimeter, cost of floor covering, and cost of baseboard. Ask the students what the most expensive scenario for the house would be, and then ask what the least expensive scenario would be. This gives a range for the total cost. You may also want to encourage students to estimate to find approximate costs before completing the table.

Encourage students to put the flooring materials and baseboards in the rooms as they would like to have it. For example, they may like tile in the bathroom instead of carpet. Encourage them to make the house as they want it, and then see if they need to make changes to come within the budget.

Rubric for Project

The following rubric can be used to assess student work.

4 The student correctly fills in the dimensions in the diagram and the total area and perimeter are correct. The student has listed three different scenarios for the house and has come within budget. All of the calculations are correct. The scenarios make sense and the student has used each material at least once. The student's work is presented neatly.

3 The student correctly fills in the dimensions in the diagram, and the total area and perimeter are correct. The student has listed three different scenarios, but may have made some minor errors in the calculations. The scenarios make sense and the student has used each material at least once. The student's work is presented neatly.

2 The student correctly fills in the dimensions in the diagram, but there are some errors in calculating the perimeter and area. The student has listed three different scenarios, but there are errors in the calculations, the scenarios do not make sense, or the student did not come within budget. The student may not have used each material at least once. The student's work is sloppy or incomplete.

1 The student does not correctly fill in the dimensions in the diagram and there are errors in calculating the perimeter and area. The student does not have three different scenarios or the scenarios are incomplete and do not make sense. The calculations are incorrect and the student did not come within the budget. The student may not have used each material at least once. The student's work is sloppy and incomplete.

CHAPTER 1

Name ______________________ Date __________

Cooperative Project: Order of Operations Game

For use after Chapter 1

Objective **To form different expressions by rearranging a set of numbers and operations as part of a game.**

Materials index cards, markers or pens, pencil, paper

Investigation *Getting Going* (Before you begin, the teacher will instruct you how to make the cards used in this game.) The game is for two players. Make two piles of cards—one pile contains the numbers zero through nine, and the other pile contains the operations $+$, $-$, $\times$, and ^{2}. Both piles should be shuffled and placed face down. Each student has one left parenthesis card and one right parenthesis card to make one set of parentheses.

Player 1 draws three numbers from the numbers pile and three operations from the operations pile. Player 1 arranges all of the number cards and as many operation cards as possible (if the ^{2} card is chosen, Player 1 can use all three operation cards chosen; if the ^{2} is not chosen, Player 1 can choose any two from the three operation cards that were drawn) to make an expression that produces the highest numerical answer. The parentheses are optional and can be used at any time. Player 1 then writes down the expression and answer on a sheet of paper.

Player 2 takes Player 1's cards. Player 2 tries to come up with an expression that produces a larger answer than Player 1. If Player 2 gets a larger answer, then Player 2 gets a point. If not, Player 1 gets a point.

Player 2 arranges the same cards to come up with an expression that produces the lowest numerical answer. Player 2 then writes down the expression and answer on a sheet of paper.

Player 1 takes the cards and tries to produce a smaller answer than Player 2. If Player 1 produces a smaller answer, then Player 1 gets a point. If Player 1 cannot come up with a smaller answer, then Player 2 gets a point. This round is over and the cards that were drawn can be put aside into a discard pile. Both players keep their parentheses cards throughout the game.

Now Player 2 draws three number cards and three operation cards. Continue the process given above. The first student to reach ten points wins. If the players run out of cards before someone has reached ten points, then the used cards from the discard pile can be shuffled and reused.

Questions

1. When trying to obtain the highest numerical result, what operation(s) do you prefer? Why?
2. When trying to obtain the lowest numerical result, what operation(s) do you prefer? Why?
3. Play the game again, but draw four numbers and four operations each time. Is this easier or harder to play than before?

Name ______________________ Date ________

Teacher's Notes for Order of Operations Project

For use after Chapter 1

Project Goals
- Create and evaluate expressions using order of operations.
- Find values of powers.

Managing the Project *Classroom Management* The students will make the cards themselves. They can work in pairs since one stack of cards will be used for two players. Index cards and markers or pens can be used to make the cards. You will need to instruct the students on how to make the cards using the following information.

Each group of two players will make thirty number cards total, three of each number using the numbers zero through nine. Make thirty operation cards total, eight of each of the operations $+$, $-$, $\times$, and six of the operation ^{2} . The operation $\div$ will not be used to avoid fraction or decimal answers and division by zero. Make four parentheses cards total, two left parentheses and two right parentheses cards. Each student will get a set of parentheses.

Instruct students to avoid negative numbers. When using the $-$ operation, students must always arrange the numbers so that the calculations produce positive numbers.

Keep in mind that you should have students with the same level of math skills play the game together. If a student that has difficulty with math is paired with a student that excels in math, the game may be unfair.

Alternative Approach A time limit could be used in this game. The student not coming up with the equation can time the student who is. Also, students can play simultaneously. For example, the students take turns drawing the numbers and operations and set the cards out so both can see them. Then they each use their own paper and pencil to come up with the equations simultaneously. They should not look at each other's work until the time is up. The student with the largest/smallest answer gets a point, and neither gets a point if they tie or come up with the same equation.

Rubric for Project **The following rubric can be used to assess student work.**

4 The students play the game correctly–valid expressions are written down and the answers given are correct. The students answer Questions 1–3 logically and insightfully. The students play the game successfully when they draw four number and four operation cards.

3 The students play the game correctly for the most part–mostly valid expressions are written and most answers are correct. The answers to Questions 1–3 are logical and somewhat insightful. The students have little difficulty playing the game as suggested in Question 3.

2 The students do not correctly play the game, but an attempt is made. Some expressions are invalid or incorrect. There are errors in the answers to the questions. The students do not correctly play the game as suggested in Question 3, but an attempt is made.

1 The students do not correctly play the game or do not attempt to. There are many errors in the answers to the questions. The students do not correctly play the game as suggested in Question 3, or they do not attempt to.

CHAPTER 1

Name ______________________________ Date __________

Independent Extra Credit Project: Patterns

For use after Chapter 1

Objective **To recognize and extend patterns.**

Materials paper, pencil, colored pencils or crayons, access to the Internet or library

Investigation *Getting Going* Consider the following pattern: 1, 3, 6, 10, 15, ... The numbers that make this pattern are called *triangular numbers*. The numbers get their name because they can be represented as dots that form triangles, as shown below. The next larger row (represented as open dots) is added to the previous triangle (represented as solid dots) to complete the pattern.

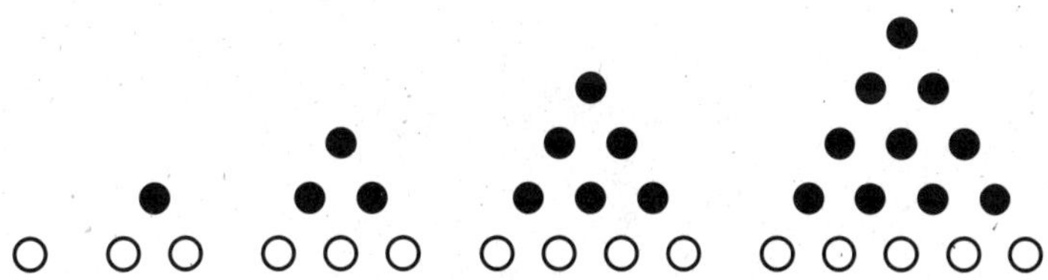

You can represent the pattern geometrically, as above, or numerically, as in the table.

Entry	Number	Pattern
1st triangular number	1	
2nd triangular number	3	1st triangular number + 2: $1 + 2 = 3$
3rd triangular number	6	2nd triangular number + 3: $3 + 3 = 6$
4th triangular number	10	3rd triangular number + 4: $6 + 4 = 10$
5th triangular number	15	4th triangular number + 5: $10 + 5 = 15$

Questions

1. Extend the pattern in the table to find the 25th triangular number.
2. Find a way to arrange dots to show your own geometric pattern. Can you represent this pattern numerically in a table?
3. Find the pattern for the *square numbers*. Use dots that form squares to represent the square numbers geometrically and your knowledge of exponents to represent the pattern numerically. Make a table of values for the first 20 square numbers.
4. Give an example of a known numeric pattern other than the triangular numbers and square numbers. You may use the Internet or another source for your search. Make a table of values for the first 15 numbers of the numeric pattern.

CHAPTER 1 Continued

Name ______________________ Date ________

Teacher's Notes for Patterns Project

For use after Chapter 1

Project Goals
- Recognizing numeric and geometric patterns.
- Extending numeric and geometric patterns.
- Writing formulas for numeric patterns.

Managing the Project *Guiding Students' Work* It may help to show the students your own example for Question 2. The pattern does not need to be complicated. It can be as easy as adding the same number of dots in a row.

To help the students with the square numbers, tell them to start with one dot. Then add to that dot the least number of dots possible to make it a square. Then add the least number of dots to make the next bigger square, and so on. Have them count the number of dots in each square —1, 4, 9, 16, etc. Help them to see the pattern numerically by asking what numbers raised to the same exponent produce these answers.

Some examples of known numeric patterns are the Fibonacci numbers and Pascal's Triangle. You may want to give one of these as an example to help with Question 4.

Rubric for Project **The following rubric can be used to assess student work.**

4 The student extends the pattern in the table and finds the 25th triangular number. The student's own pattern is creative, complete, and correct, and is represented both geometrically and numerically. The student correctly illustrates the square numbers and correctly completes a table of values for the first 20 square numbers. The student gives an example of a known numeric pattern. A table of values for the first 15 numbers of the numeric pattern is complete and correct. The student's work is neat.

3 The student extends the pattern in the table and finds the 25th triangular number. The student's own pattern is creative and mostly correct and is represented geometrically and numerically with almost no errors. The student illustrates the square numbers and completes a table of values for the first 20 but may have slight errors. The student gives an example of a known numeric pattern. A table of values for the first 15 numbers of the numeric pattern is complete and mostly correct. The student's work is neat.

2 The student may have some errors in the table and the 25th triangular number may be incorrect. The student's own pattern has some errors and is not represented correctly geometrically and/or numerically. The student does not correctly illustrate the square numbers and the table for the first 20 has errors. The student's example of a known numeric pattern is incorrect or does not make sense and the table of values was attempted but has errors. The work is sloppy or incomplete.

1 There are errors in the table and the 25th triangular number is incorrect. The student's own pattern is incomplete or incorrect and is not correctly represented. The student does not correctly illustrate the square numbers and the table is incomplete or incorrect. The student did not find a known pattern or it is not correct and the table is incomplete or incorrect. The work is sloppy and incomplete.

CHAPTER 1

Name ______________________ Date __________

Cumulative Practice

For use after Chapter 1

Describe the pattern. Then write the next three numbers. (Lesson 1.1)

1. 11, 21, 31, 41, . . .

2. 55, 50, 45, 40, . . .

3. 3, 9, 27, 81, . . .

4. 64, 32, 16, 8, . . .

Evaluate the expression when $a = 12$ and $b = 5$. (Lesson 1.2)

5. $a - b$

6. $3b + 4$

7. ab

Write the product as a power. (Lesson 1.3)

8. $2 \cdot 2 \cdot 2 \cdot 2 \cdot 2$

9. $15 \cdot 15 \cdot 15$

10. $z \cdot z$

Evaluate the power. (Lesson 1.3)

11. 10^3

12. 3^4

13. 7^3

Evaluate the expression. (Lesson 1.4)

14. $44 - 6 \cdot 4$

15. $\frac{20 - 8}{2 \cdot 3}$

16. $36 - (4 + 2)$

17. $25 - 4^2 \div 8$

Evaluate the expression when $x = 10$, $y = 16$, and $z = 2$. (Lessons 1.2, 1.3, and 1.4)

18. $2x - (y + 3)$

19. $x^2 + z^3$

20. $\frac{x + y}{y - 7z}$

21. $x + y \div z^2$

Tell whether the given value of the variable is a solution of the equation. (Lesson 1.5)

22. $4 + t = 15; t = 12$

23. $\frac{38}{y} = 2; y = 19$

24. $12x = 84; x = 7$

25. $k - 8 = 5; k = 3$

Use the formula for distance to find the unknown value. (Lesson 1.5)

26. $d = 140$ miles, $r = 35$ miles per hour, $t =$ __?__

27. $d = 210$ meters, $r =$ __?__ , $t = 30$ seconds

Review and Projects

CHAPTER 1 Continued

Name ______________________ Date ________

Cumulative Practice

For use after Chapter 1

28. A car moves at a speed of 55 miles per hour. At this speed, how long would it take to drive 165 miles? (Lesson 1.5)

Solve the equation using mental math. (Lesson 1.5)

29. $g - 11 = 20$

30. $22 + n = 29$

31. $46 = 2m$

32. $37 \div z = 1$

33. $p - 0 = 100$

34. $32 - x = 20$

Find the perimeter and the area of the rectangle or square with the given dimensions. (Lesson 1.6)

35.

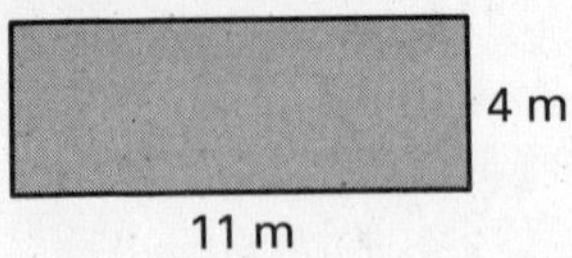

36.

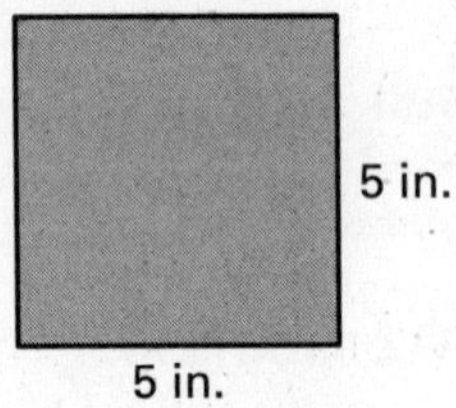

37. $s = 14$ feet

38. $\ell = 21$ yards, $w = 9$ yards

39. Which pizza has greater area: a rectangular pizza with a length of 24 inches and a width of 18 inches, or a square pizza with a side length of 21 inches? (Lesson 1.6)

Use the strategy *look for a pattern*. (Lesson 1.7)

40. Study the table for the powers of 6. Without actually calculating 6^{11}, how can you tell what its tens digit is?

Power	6^1	6^2	6^3	6^4	6^5	6^6	6^7	6^8
Value	6	36	216	1296	7776	46,656	279,936	1,679,616

41. You keep track of how long it takes to complete your math homework each night. For the first several nights the number of minutes are 15, 30, 19, 30, 23, and 30. How long would you expect to spend on math homework on the ninth night?

Use any strategy to solve the problem. (Lesson 1.7)

42. A parking meter accepts only quarters, dimes, and nickels. How many different coin combinations could be used to pay the $.50 charge for an hour of parking?

Review and Projects

Answers

Lesson 1.1

Activity Master

1. *Sample Answer:* Each figure has two more squares than the previous figure. **2.** The 7th figure will have 13 squares and the 8th figure will have 15 squares, both arranged in the shape of an "L".

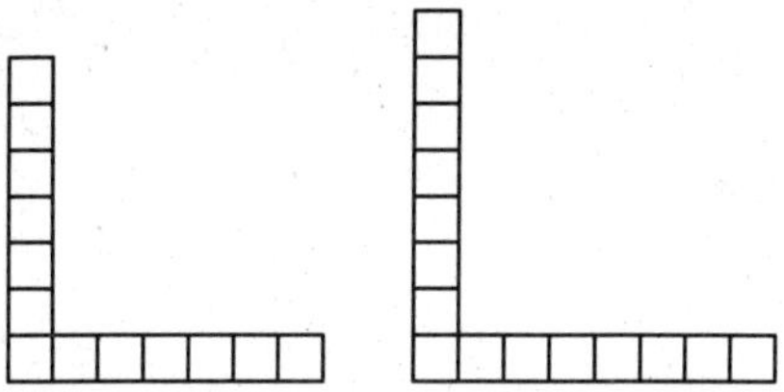

3. The next three symbols will be a circle, a square, and a circle. 

4. Answers will vary.

Practice A

1. 863 **2.** 341 **3.** 3976 **4.** 55 **5.** D

6. B **7.** A **8.** C **9.** Add 2 to the previous number. Next three numbers: 9, 11, 13

10. Subtract 4 from the previous number. Next three numbers: 28, 24, 20 **11.** Multiply the previous number by 8. Next three numbers: 4096, 32,768, 262,144 **12.** Add a column of two dots to the previous figure. Next figure: 

13. If the previous figure is a square, add a triangle. If the previous figure is a triangle, add a square. Next figure: 

14. 5000 miles; 10,000 miles; 15,000 miles; 20,000 miles; 25,000 miles; Add 5000 to the previous number of miles. **15.** Add 7 to the previous number. Next three numbers: 31, 38, 45

16. Divide the previous number by 3. Next three numbers: 9, 3, 1 **17.** Subtract 3 from the previous number. Next three numbers: 17, 14, 11

18. Turn the previous figure one-quarter turn to the right. Next figure: 

19. Turn the previous figure one-quarter turn to the left. Next figure: 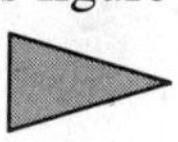

20. 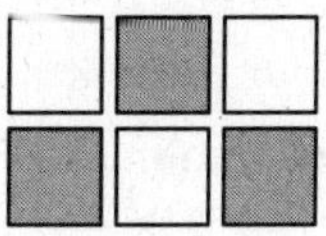

21. You will run for 35 minutes during the ninth week.

Practice B

1. Multiply the previous number by 3. Next three numbers: 81, 243, 729 **2.** Add 6 to the previous number. Next three numbers: 28, 34, 40

3. Subtract 2 from the previous number. Next three numbers: 13, 11, 9 **4.** Multiply the previous number by 4. Next three numbers: 768, 3072, 12,288 **5.** Divide the previous number by 2. Next three numbers: 400, 200, 100

6. Multiply the previous number by 5. Next three numbers: 2500, 12,500, 62,500 **7.** The next four leap years are in 2004, 2008, 2012, and 2016.

8. Add a row of blocks to the previous figure that is one block longer than the bottom row of the previous figure. Next figure: 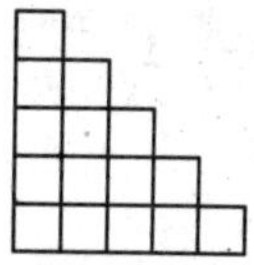

9. Turn the previous figure one-sixth of a turn to the right. Next figure: 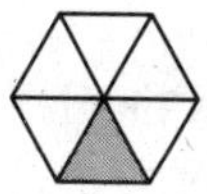

10. 3750 miles; 7500 miles; 11,250 miles; 15,000 miles; 18,750 miles; Add 3750 miles to the previous number of miles. **11.** Each letter in the pattern is the third letter after the previous letter. Next three letters: M, P, S **12.** Starting with the last letter in the alphabet, each letter in the pattern is the second letter before the previous letter. Next three letters: R, P, N **13.** Each letter in the pattern is the fourth letter after the previous letter. Next three letters: Q, U, Y

14. Turn the previous figure one-fifth of a turn to the left. Next figure: 

15. If the previous figure is a circle, draw a shaded square that encloses the circle. If the previous figure is a square, draw a circle that encloses the square. Next figure. 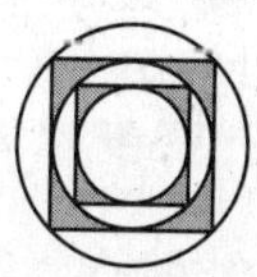

16. The year of the horse occurs again in 1990, 2002, 2014, 2026, and 2038. Add 12 to the previous year.

Lesson 1.1 *continued*

17. Next three figures in the pattern:

Practice C

1. Multiply the previous number by 4. Next three numbers: 256, 1024, 4096 **2.** Add 11 to the previous number. Next three numbers: 46, 57, 68

3. Subtract 8 from the previous number. Next three numbers: 42, 34, 26 **4.** Multiply the previous number by 3. Next three numbers: 486, 1458, 4374 **5.** Divide the previous number by 2. Next three numbers: 112, 56, 28 **6.** Subtract 6 from the previous number. Next three numbers: 67, 61, 55 **7.** First four numbers in pattern: 5, 35, 245, 1715

8. Next two meetings: April 15, April 29

9. You describe and extend a numerical pattern by first trying to find a relationship between the first and second numbers. Then check to see if the relationship is true for the second and third numbers, the third and fourth numbers, and so on.

10. The next rectangle's length is one dot longer than the previous rectangle's length.
Next rectangle pattern: 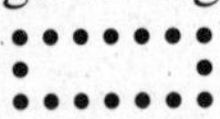

11. The next square's length and width are each divided into a number of equal segments that are one more than the number of equal segments used for the previous square's length and width. Next figure in pattern: 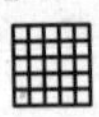

12. Next five fish in pattern:

13.

14. They will appear at the same time in 2013. To find the answer to the problem, you write the pattern for both groups of cicadas for a number of years and see when the same year occurs in both lists.

Study Guide

1. 28, 33, 38, 43, 48 **2.** 8:28 **3.** Add 8 to the previous number to get the next number in the pattern; 46, 54, 62 **4.** Subtract 9 from the previous number to get the next number in the pattern; 31, 22, 13 **5.** Multiply the previous number by 4 to get the next number in the pattern; 1024, 4096, 16,384 **6.** Divide the previous number by 3 to get the next number in the pattern; 9, 3, 1 **7.** 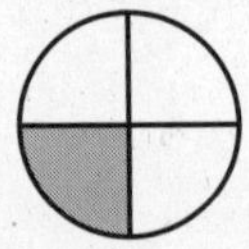 **8.** 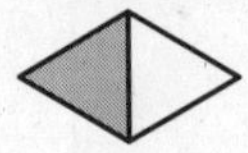

Challenge Practice

1. Multiply the previous number by 6; 16,848, 101,088, 606,528 **2.** Subtract 111 from the previous number; 234, 123, 12 **3.** Divide the previous number by 2; $\frac{1}{8}$, $\frac{1}{16}$, $\frac{1}{32}$ **4.** Alternate: add 10 to the previous number, subtract 4 from the previous number; 19, 29, 25 **5.** Add the previous number; 43, 69, 112 **6.** If even then divide the previous number by 2, if odd then add 5 to the previous number; 4, 2, 1 **7.** 4, 8, 12, 16; Add 4 to the previous number; 20, 24, 28 **8.** 1, 4, 9, 16; add 3, add 5, add 7, . . . ; 25, 36, 49

Lesson 1.2

Technology Activity

1. 83 **2.** 88 **3.** 126 **4.** 110 **5.** 128 **6.** 96 **7.** 4 **8.** 12 **9.** 180 **10.** 315 **11.** yes **12.** no **13.** 13 **14.** 21

Practice A

1. 3 **2.** 10 **3.** 12 **4.** 15 **5.** 6 **6.** 9 **7.** evaluate **8.** 6 **9.** 6 **10.** 16 **11.** 2 **12.** 1 **13.** 3 **14.** The total cost for the month is $60. **15.** 21 **16.** 5 **17.** 84 **18.** 40 **19.** 26 **20.** 136 **21.** You will have to pay $17 for the jeans. **22.** It will take you 2 hours to get to your relative's house.

Practice B

1. 21 **2.** 17 **3.** 91 **4.** 15 **5.** 29 **6.** 27 **7.** You will earn $105 during your first week. **8.** There are 48 calories from protein in the serving. **9.** 43 **10.** 72 **11.** 11 **12.** 3

Lesson 1.2 *continued*

13. 16 **14.** 38 **15.** It feels 19 degrees colder because of the wind. **16.** It will take you 8 hours to drive across the state. **17.** a and c

Practice C

1. 78 **2.** 38 **3.** 462 **4.** 9 **5.** 144 **6.** 6 **7.** You will have to pay $161 for repairs. **8.** 56 **9.** The service fee is $7. **10.** 56 **11.** 29 **12.** 1050 **13.** 5 **14.** 210 **15.** 27 **16.** The total cost for lunch is $77. **17.** The symbol × can be confused for the variable *x*. **18.** 140 **19.** 240 **20.** 21 **21.** d

Study Guide

1. 12 **2.** 7 **3.** 23 **4.** 13 **5.** 20 **6.** 1 **7.** 40 **8.** 12 **9.** 48 **10.** 7 **11.** 2 **12.** 60 **13.** 12 **14.** 3 **15.** 28 **16.** 12 **17.** 8 **18.** 9

Real World Problem Solving

1. Class 1: 64 folders; Class 2: 60 folders; Class 3: 58 folders; Class 4: 66 folders **2.** 120 **3.** yes

Challenge Practice

1. $\frac{8}{3}$ **2.** 18 **3.** 0 **4.** 100 **5.** 107 **6.** 97 **7.** 9 **8.** 4

Lesson 1.3

Practice A

1. 16 **2.** 49 **3.** 144 **4.** power **5.** B **6.** C **7.** D **8.** A **9.** 8^3 **10.** 3^4 **11.** 6^5 **12.** y^3 **13.** m^6 **14.** n^2 **15.** 8 **16.** 81 **17.** 125 **18.** 49 **19.** 81 **20.** 64 **21.** There are 100 centimeters in one meter. **22.** There are 2^3 bits in one byte. **23.** The rug is made up of 5^2, or 25, squares.

Practice B

1. three to the eighth power **2.** nine to the fourth power **3.** *y* to the fifth power **4.** 2^6 **5.** 13^3 **6.** 6^4 **7.** r^3 **8.** a^7 **9.** t^4 **10.** 32 **11.** 2401 **12.** 1000 **13.** 243 **14.** 216 **15.** 1 **16.** 144 **17.** 1024 **18.** 3375 **19.** 729 **20.** 10,000 **21.** 256 **22.** 5^2 **23.** 11^2 **24.** 8^2 or 4^3 or 2^6 **25.** $1^2, 2^2, 3^2, 4^2, 5^2, \ldots$ **26.** There are 1024 bytes in a kilobyte. **27.** There are 10^3 milligrams in one gram. **28.** There are 3^3, or 27, blocks in the cube.

Practice C

1. seven to the eleventh power **2.** *m* to the fourth power **3.** *t* to the twelfth power **4.** 1^4 **5.** 4^5 **6.** t^9 **7.** 7776 **8.** 512 **9.** 1331 **10.** 0 **11.** 1 **12.** 3125 **13.** 24 **14.** 4096 **15.** 4096 **16.** 15,625 **17.** 160,000 **18.** 1 **19.** 25^2 or 5^4 **20.** 20^2 **21.** 1^1 **22.** 2^5 **23.** 5^3 **24.** 3^3 **25.** $10^1 = 10$, $10^2 = 100$, $10^3 = 1000$, $10^4 = 10{,}000$. The number of the exponent is the same as the number of zeros in the evaluated power. $10^5 = 100{,}000$, $10^6 = 1{,}000{,}000$, and $10^7 = 10{,}000{,}000$. **26.** There are 1,048,576 bytes in a megabyte. **27.** There are 4^3, or 64, blocks in the cube. **28.** 64 **29.** 81 **30.** 25

Study Guide

1. 10^3 **2.** z^7 **3.** 8^5 **4.** v^2 **5.** 1024 **6.** 512 **7.** 5 **8.** 1024 **9.** 7776

Challenge Practice

1. 10^5 **2.** $(x + 1)^4$ **3.** 117 **4.** 387 **5.** 141 **6.** 263 **7.** $2^4; 4^2$ **8.** $3^4; 9^2$ **9.** $3^6; 9^3; 27^2$ **10.** $10^6; 100^3; 1000^2$

Lesson 1.4

Practice A

1. 16 **2.** 27 **3.** 100 **4.** fraction bars **5.** multiplication **6.** addition **7.** addition **8.** multiplication **9.** addition **10.** division **11.** 36 **12.** 11 **13.** 10 **14.** 78 **15.** 16 **16.** 4 **17.** 43 **18.** 11 **19.** 6 **20.** 15 **21.** 8 **22.** 15 **23.** 5837 **24.** $50 + 45 \cdot 12 = 590$; It will cost $590 to get cable for the first year. **25.** $40 \cdot 5 + 20 \cdot 3 = 260$; It will cost $260 for the group to enter the park.

Lesson 1.4 *continued*

Practice B

1. 58 **2.** 31 **3.** 90 **4.** 4 **5.** 8 **6.** 3 **7.** 720 **8.** 11 **9.** 172 **10.** 137 **11.** 60 **12.** 15 **13.** 320 **14.** 1 **15.** 78 **16.** 4 **17.** 52 **18.** 0 **19.** B **20.** A **21.** C

22. $100 + 40 \cdot 12 = 580$; DSL service will cost $580 for the first year.

23. $50 \cdot 1 + 30 \cdot 5 + 8 \cdot 10 = 280$; There is $280 in the cash register.

Practice C

1. 2 **2.** 48 **3.** 6 **4.** 4 **5.** 540 **6.** 6 **7.** 57 **8.** 3 **9.** 4 **10.** 115 **11.** 103 **12.** 2 **13.** 181 **14.** 5 **15.** 129 **16.** 1 **17.** 0 **18.** 914 **19.** $11 + 3 \cdot (17 - 5)$

20. $6 \cdot 6 + 1 \cdot 33 + 3 \cdot 18 = 123$;
The place-kicker will score 123 points for the season. **21.** b and c

Study Guide

1. $45 + 20 \times 12$; $285 **2.** 2 **3.** 32 **4.** 23 **5.** 26 **6.** 2 **7.** 38 **8.** 87 **9.** 80 **10.** 3 **11.** 19

Challenge Practice

1. 12 **2.** 7 **3.** 56 **4.** 20 **5.** $-, \div$ **6.** $\div, \cdot$ **7.** $+, \div, -$ **8.** $+, \div, -, \cdot$

Lesson 1.5

Practice A

1. 9 **2.** 3 **3.** 7 **4.** travel time **5.** B **6.** A **7.** C **8.** D **9.** 6 is a solution of the equation. **10.** 12 is not a solution of the equation. **11.** 28 is a solution of the equation.

12. 9 is not a solution of the equation.

13. 12 is a solution of the equation.

14. 28 is not a solution of the equation. **15.** 6 **16.** 12 **17.** 4 **18.** 5 **19.** 8 **20.** 4

21. Bob Hayes' running speed was 10 meters per second. **22.** It would take you 3 hours to bike 30 miles. **23.** You need to work 20 hours to meet your goal. **24.** The temperature is 100 degrees Celsius.

Practice B

1. 6 is a solution of the equation. **2.** 12 is a solution of the equation. **3.** 2 is not a solution of the equation. **4.** 34 is not a solution of the equation. **5.** 13 is a solution of the equation.

6. 25 is not a solution of the equation. **7.** C; 8 **8.** B; 48 **9.** A; 42 **10.** D; 8 **11.** 8 **12.** 9 **13.** 5 **14.** 8 **15.** 40 **16.** 19 **17.** 20 miles per hour **18.** 9 hours **19.** 60 miles per hour

20. It takes the bird 3 hours to fly 15 miles.

21. The focal length of the telescope is 90 inches. **22.** The temperature is 27 degrees Celsius.

Practice C

1. 16 is a solution of the equation. **2.** 13 is not a solution of the equation. **3.** 13 is a solution of the equation. **4.** 32 is not a solution of the equation. **5.** 26 is a solution of the equation.

6. 7 is not a solution of the equation. **7.** 9 **8.** 11 **9.** 19 **10.** 6 **11.** 15 **12.** 32

13. 4 hours **14.** 41 miles per hour

15. 20 hours **16.** It takes sound 4 seconds to travel through 12 miles of iron.

17. The focal length of the telescope is 150 inches. **18.** The temperature is 177 degrees Celsius. **19.** $96 \div x = 32$; $x = 3$

20. 15 **21.** 5 **22.** 6

Study Guide

1. yes **2.** no **3.** yes **4.** 20 **5.** 6 **6.** 8 **7.** 27 **8.** 7 min

Challenge Practice

1. 22 **2.** 16 **3.** 0 **4.** 3 **5.** 49 **6.** 16 **7.** 7 **8.** 9 **9.** Your brother arrives 200 seconds $\left(3\frac{1}{3}\text{ minutes}\right)$ after you.

Lesson 1.6

Practice A

1. 9 **2.** 100 **3.** 64 **4.** product **5.** C

6. Perimeter = 20 centimeters; Area = 16 square centimeters

7. Perimeter = 20 feet; Area = 25 square feet

Lesson 1.6 *continued*

8. Perimeter = 26 yards; Area = 42 square yards

9. Perimeter = 36 meters; Area = 81 square meters **10.** The perimeter of the computer screen is 46 inches and the area of the computer screen is 130 square inches. **11.** The perimeter of the photograph is 24 inches and the area of the photograph is 35 square inches. **12.** The perimeter of the flower bed is 34 feet and the area of the flower bed is 60 square feet. **13.** The perimeter of the tablet is 76 feet and the area of the tablet is 336 square feet.

Practice B

1. B **2.** C **3.** D **4.** A **5.** Perimeter = 32 inches; Area = 55 square inches

6. Perimeter = 16 meters; Area = 16 square meters **7.** Perimeter = 40 millimeters; Area = 100 square millimeters

8. Perimeter = 30 feet; Area = 26 square feet

9. Perimeter = 30 centimeters; Area = 56 square centimeters **10.** Perimeter = 28 inches; Area = 49 square inches **11.** Perimeter = 12 meters; Area = 9 square meters

12. Perimeter = 36 yards; Area = 72 square yards

13. A and D **14.** The perimeter of the screen is about 204 millimeters and the area of the screen is about 2501 square millimeters. **15.** The perimeter of the rectangle is 12 centimeters and the area of the rectangle is 8 square centimeters.

Practice C

1. Perimeter = 42 inches; Area = 108 square inches **2.** Perimeter = 32 meters; Area = 64 square meters **3.** Perimeter = 24 millimeters; Area = 36 square millimeters

4. Perimeter = 46 feet; Area = 132 square feet

5. Perimeter = 100 centimeters; Area = 600 square centimeters **6.** Perimeter = 60 inches; Area = 225 square inches

7. Perimeter = 56 meters; Area = 196 square meters **8.** Perimeter = 46 yards; Area = 120 square yards **9.** B and C

10. The perimeter of the field is 1040 feet and the area of the field is 57,600 square feet.

11. The difference in area between the screens is 20 square inches.

12. ℓ = 4 feet, w = 4 feet; ℓ = 8 feet, w = 2 feet; ℓ = 16 feet, w = 1 foot **13.** The perimeter of the rectangle is 40 inches and the area of the rectangle is 99 square inches. **14.** The perimeter of the figure is 42 feet and the area of the figure is 40 square feet. First you need to find the missing dimensions of the figure to find the perimeter. Then you break the figure into three rectangles as shown below to find the area.

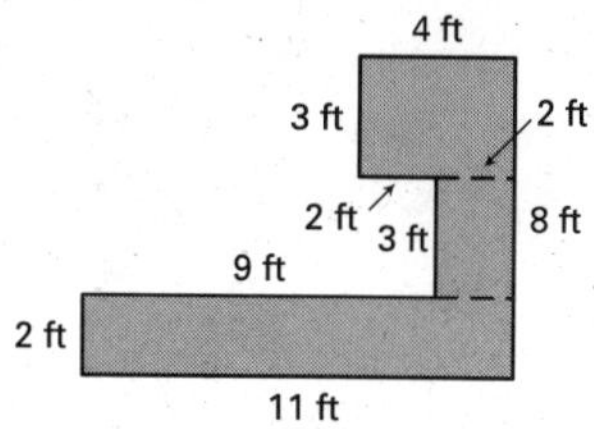

Study Guide

1. 16 ft **2.** 16 cm **3.** 32 m^2 **4.** 49 yd^2 **5.** 30 cm; 54 cm^2 **6.** 20 in.; 24 $in.^2$

Real World Problem Solving

1. Room 1: 108 ft^2; Room 2: 140 ft^2; Room 3: 132 ft^2; Room 4: 120 ft^2; Room 5: 126 ft^2

2. Driveway: 900 ft^2 **3.** Backyard: 200 ft

4. 160 ft

Challenge Practice

1. Perimeter = 28 centimeters; Area = 22 square centimeters **2.** Area = 1296 square meters

3. Perimeter = 48 meters

4. 16 inches; Area = 256 square inches

5. Answers may vary.
Sample Answer:

Length	Width
60 feet	2 feet
40 feet	3 feet
30 feet	4 feet
15 feet	8 feet
12 feet	10 feet

6. 5 feet **7.** length = 20 feet; width = 6 feet

Lesson 1.7

Practice A

1. \$25 **2.** \$75 **3.** You could buy the tan pants and the striped shirt, the tan pants and the solid shirt, or the navy pants and the solid shirt.

4. division **5.** addition **6.** multiplication **7.** subtraction **8.** 35 **9.** 264 square feet

Practice B

1. The remainder wasn't taken into consideration, so there's only enough pizza for 32 people. You need to order 5 pizzas. **2.** It will take 7 seasons to return to the original positions of the crops. **3.** You could buy the navy pants and the striped shirt, the navy pants and the solid shirt, or the black pants and the solid shirt.

4. There are 307,200 pixels with the 640 × 480 resolution and there are 480,000 pixels with the 800 × 600 resolution. There are 172,800 more pixels with the 800 × 600 resolution.

5. 10; ROP, ROS, ROM, RPS, RPM, RSM, OPS, OPM, OSM, PSM

Practice C

1. If your answer is not reasonable, you check to see if you set up the problem incorrectly then check your solution to see if you made any arithmetic errors.

2.

Power	7^1	7^2	7^3
Value	7	49	343

Power	7^4	7^5	7^6
Value	2401	16,807	117,649

There appears to be a pattern in the last digits of the powers of 7 that repeats: 7, 9, 3, 1, 7, 9, 3, 1, ... So the last digit of 7^7 should be 3. **3.** You could buy the tan pants and the solid shirt, the navy pants and any one of the shirts, or the black pants and any one of the shirts.

4. There are 307,200 pixels with the 640 × 480 resolution and there are 786,432 pixels with the 1024 × 768 resolution. There are 479,232 more pixels with the 1024 × 768 resolution.

5. The total area of the pasture is 800 square fe and the farmer needs 140 feet of fencing.

6. Judy is 63 inches tall, Keith is 64 inches tall, and Sara is 65 inches tall.

Study Guide

1. \$42 **2.** \$158 **3.** amusement park, pool, zoo; amusement park, pool, basketball camp; amusement park, pool, camping

Challenge Practice

1. 9 **2.** 11 sheep; 8 chickens **3.** 833

4. 15 yards by 15 yards **5.** 9 minutes

6. You arrive 2 minutes before your sister.

Review and Projects

Chapter Review Games and Activities

1. A **2.** B **3.** A **4.** B **5.** A **6.** B **7.** A **8.** B **9.** A

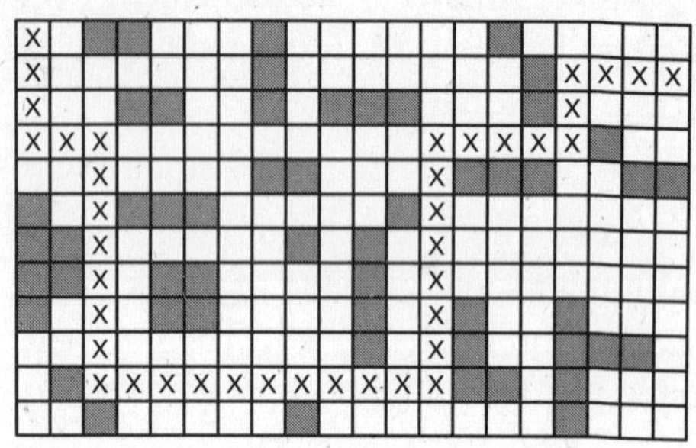

Real-Life Project

1. Total area = 960 square feet
Total perimeter = 124 feet

2. Check student work.

Cooperative Project

1–3. Answers may vary.

Independent Extra Credit Project

1.

Entry	Number	Pattern
1st	1	
2nd	3	1st + 2: 1 + 2 = 3
3rd	6	2nd + 3: 3 + 3 = 6
4th	10	3rd + 4: 6 + 4 = 10

Review and Projects *continued*

Entry	Number	Pattern
5th	15	4th + 5: 10 + 5 = 15
6th	21	5th + 6: 15 + 6 = 21
7th	28	6th + 7: 21 + 7 = 28
8th	36	7th + 8: 28 + 8 = 36
9th	45	8th + 9: 36 + 9 = 45
10th	55	9th + 10: 45 + 10 = 55
11th	66	10th + 11: 55 + 11 = 66
12th	78	11th + 12: 66 + 12 = 78
13th	91	12th + 13: 78 + 13 = 91
14th	105	13th + 14: 91 + 14 = 105
15th	120	14th + 15: 105 + 15 = 120
16th	136	15th + 16: 120 + 16 = 136
17th	153	16th + 17: 136 + 17 = 153
18th	171	17th + 18: 153 + 18 = 171
19th	190	18th + 19: 171 + 19 = 190
20th	210	19th + 20: 190 + 20 = 210
21st	231	20th + 21: 210 + 21 = 231
22nd	253	21st + 22: 231 + 22 = 253
23rd	276	22nd + 23: 253 + 23 = 276
24th	300	23rd + 24: 276 + 24 = 300
25th	325	24th + 25: 300 + 25 = 325

2. Answers may vary.

3.

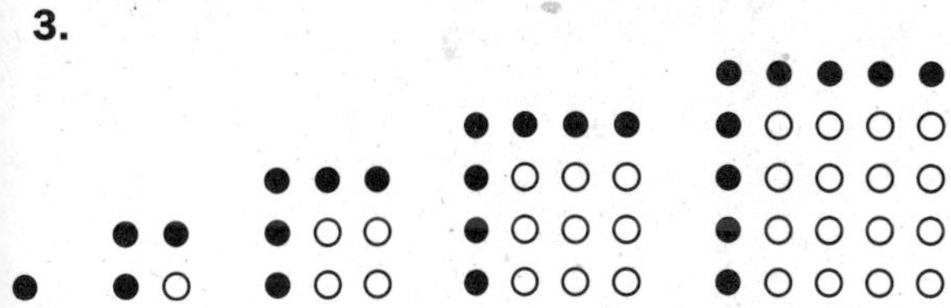

Entry	Number	Pattern
1st square number	1	1^2
2nd square number	4	2^2
3rd square number	9	3^2
4th square number	16	4^2
5th square number	25	5^2
6th square number	36	6^2
7th square number	49	7^2
8th square number	64	8^2
9th square number	81	9^2
10th square number	100	10^2
11th square number	121	11^2
12th square number	144	12^2
13th square number	169	13^2
14th square number	196	14^2
15th square number	225	15^2
16th square number	256	16^2
17th square number	289	17^2
18th square number	324	18^2
19th square number	361	19^2
20th square number	400	20^2

4. Answers may vary.

Cumulative Practice

1. Add 10 to the previous number; 51, 61, 71

2. Subtract 5 from the previous number; 35, 30, 25 **3.** Multiply the previous number by 3; 243, 729, 2187 **4.** Divide the previous number by 2; 4, 2, 1 **5.** 7 **6.** 19 **7.** 60

8. 2^5 **9.** 15^3 **10.** z^2 **11.** 1000 **12.** 81

13. 343 **14.** 20 **15.** 2 **16.** 30 **17.** 23

18. 1 **19.** 108 **20.** 13 **21.** 14 **22.** no

23. yes **24.** yes **25.** no **26.** 4 hours

27. 7 meters per second **28.** 3 hours **29.** 31

30. 7 **31.** 23 **32.** 37 **33.** 100 **34.** 12

35. Perimeter = 30 meters; Area = 44 square meters **36.** Perimeter = 20 inches; Area = 25 square inches **37.** Perimeter = 56 feet; Area = 196 square feet **38.** Perimeter = 60 yards; Area = 189 square yards

39. square pizza **40.** 5 **41.** 31 minutes

42. 10 different combinations